Mobile Methodologies

Mobile Methodologies

Edited By

Ben Fincham
University of Sussex, UK

Mark McGuinness
Bath Spa University, UK

Lesley Murray
University of Brighton, UK

Foreword by Mimi Sheller
Drexel University, USA

palgrave
macmillan

Selection and editorial matter © Ben Fincham, Mark McGuinness and
Lesley Murray 2010
Individual chapters © their respective authors 2010
Foreword © Mimi Sheller 2010

First published 2010 by
PALGRAVE MACMILLAN

Palgrave Macmillan in the UK is an imprint of Macmillan Publishers Limited,
registered in England, company number 785998, of Houndmills, Basingstoke,
Hampshire RG21 6XS.

Palgrave Macmillan in the US is a division of St Martin's Press LLC,
175 Fifth Avenue, New York, NY 10010.

Palgrave Macmillan is the global academic imprint of the above companies
and has companies and representatives throughout the world.

Palgrave® and Macmillan® are registered trademarks in the United States,
the United Kingdom, Europe and other countries.

ISBN 978-0-230-59442-5   hardback

This book is printed on paper suitable for recycling and made from fully
managed and sustained forest sources. Logging, pulping and manufacturing
processes are expected to conform to the environmental regulations of the
country of origin.

A catalogue record for this book is available from the British Library.

A catalogue record for this book is available from the Library of Congress.

10   9   8   7   6   5   4   3   2   1
19 18 17 16 15 14 13 12 11 10

Printed and bound in Great Britain by
CPI Antony Rowe, Chippenham and Eastbourne

Contents

Foreword

Over the last decade, inspired by the mobility turn in the social sciences, numerous researchers have been out in the field exploring various ways of researching and representing mobilities in all of their messiness and complexity. Yet the methodological implications of mobilities research are just beginning to be unpacked and critically evaluated. It is now nearly six years since the Centre for Mobilities Research (CeMoRe) hosted the Alternative Mobility Futures conference at Lancaster University (January, 2004), along with a Mobile Communications Workshop, which together resulted in two edited collections (Sheller & Urry, 2006a, 2006b) and inspired the creation of the journal *Mobilities*. The article 'The New Mobilities Paradigm', which opened the special issue of *Environment and Planning A* on Materialities and Mobilities, summed up the developments of this 'mobility turn' and called for new research methods that would be 'on the move' and would 'simulate intermittent mobility' (Sheller & Urry, 2006c: 217).

The methods called for then included: interactional and conversational analysis of people as they moved; mobile ethnography involving itinerant movement with people, following objects, and co-present immersion in various modes of movement; after the fact interviews and focus groups about mobility; the keeping of textual, pictorial or digital time-space diaries; various methods of cyber-research, cyber-ethnography and computer simulations; imaginative travel using multimedia methods attentive to the affective and atmospheric feeling of place; the tracking of affective objects that attach memories to place and circulate in 'sticky' ways; and finally methods that measure the spatial structuring and temporal pulse of transfer points and places of in-between-ness in which the circulation of people and objects are slowed or stopped, as well as facilitated and speeded. This laid the groundwork for Urry's chapter on mobile methods in his book *Mobilities* (2007) and the more recent exploration of mobile methods and the empirical (Buscher & Urry, 2009) that arose out of a CeMoRe workshop on Mobile Methods (February, 2008).

In line with these empirical investigations of diverse mobilities, in *Mobile Methodologies* one can see the outcomes of new research by those who have been inspired by the mobility turn as well as those who have more critically engaged with it. This is one of the first collections to interrogate diverse methodologies for mobilities research and to con-sider problems such as the limits to representationality, dilemmas in research ethics, and the epistemological challenges concerning what we consider mobility to be in the first place. Mobilities research in its broadest sense concerns not only physical movement, but also potential movement, blocked movement, immobilisation, and forms of dwelling and place-making (Buscher & Urry, 2009). In re-engaging the empirical and exploring the relation between observation and participation, this collection emphasises methods not only for observ-ing movement, but also for 'investigating placed narratives' (Murray), discovering tacit knowledge and what is taken-for-granted (Freudendal-Pedersen, Hartmann-Petersen and Drewes Nielsen), learning to 'incorpo-rate movement as a methodological tool and a means of participation' (Lashua and Cohen), and 'actively engaging movement in the creation of research knowledge' (Brown and Spinney).

Delving into rich empirical data, including the occasional banality (and humour) of mundane practices of mobility, the chapters included here emphasise the use of qualitative data, mixed methods, and participatory reflexivity, in contrast to the discourses of scientific objectivity, quan-tification, and rational choice that are usually deployed in transport geo-graphy and in wider public policy debates about transport. This approach opens up mobilities research to encompass not only studies of walking, driving, bicycling, passengering and using public transport, all of which are represented here, but also to explore the kinaesthetic and sensory aspects of musicscapes, dance or martial arts, transnational travel, and auto-biographies of movement. This wide palette of research interests challenges the limitations of narrow conceptions of mobilities as 'move-ment from A to B' (cf. Cresswell, 2006). It gets inside the embodied experience and accomplishment of movement, and surprisingly demands that research also attend to the less animated 'quiescent sensibilities' of more 'passive mobilities' (Bissell, Ch. 4). Stilling, re-tracing, re-screening, narrating, and making meaning all move to the fore.

The volume reflects the turn towards what Clarke (Ch. 8) calls 'a more mature mobility studies', which in contrast to earlier 'nomadic theory' has attempted to capture the dialectics of mobility and immobility, to analyse the relations of power that shape the meanings and practices of mobility, and to acknowledge the more ephemeral, embodied, and affec-tive dimensions of interlocking relational (im)mobilities (see e.g. Hannam *et al.*, 2006; Cresswell, 2006). Here in *Mobile Methodologies* one sees deeply engaged empirical researchers reflect on their own practice, highlight the methods that they have used and those that have failed them, and push the limits of technologies such as video-recording, participant-observation on-the move, and combinations of these methods with interviewing, action research, focus groups and narration. The contributors reflect on the difficulties of these methodologies as much as the successes, the problems in translation from the non-representational to the representational, and the power relations and distortions that may be inherent in the research process itself.

The contributors cover many methods, while leaving others still to be worked on, but undoubtedly in a context informed by the kind of epistemological and methodological concerns raised here. Walker's discussion of covert research on 'naïve drivers' (Ch. 3), for example, can be extended to thinking about the use of other forms of covert surveillance and tracking that are offered to researchers by new kinds of mobile informatics and geolocational devices. How can self-reported and self-narrated mobility be combined not only with video recording and interviewing, but also with geographical positioning systems (GPS), mobile geographical information systems (GIS), or simply use of mobile phones? What are the ethical implications of this kind of multi-modal research and how can informed consent be given? Does it matter whether research subjects are children, bus drivers, working-class women, or others in unequal power relations with the researcher, and how can that be dealt with specifically within the field of mobilities research as these new mixed methods come into play?

Other chapters raise issues about participatory research, including engagement in potentially dangerous activities, questions of self-immersion and auto/biography, and aspects of action research and elicitation of reflexivity. These are all crucial issues for mobilities research methodologies, as researchers 'enter the field' in new ways, collect data on-the-move, and reflect on the meaning of mobilities along with their research participants as they carry out the research. Those who are about to engage in their own research or who are teaching research methodologies can certainly learn much from reflecting on the issues and examples presented here. Beyond the practical matters of carrying out the research, the chapters also address underlying fundamental questions about the ontological status of physical bodies, virtual bodies, or simulated bodies, and the relations achieved across physical, virtual, or simulated spaces. Although the preponderant scale of interest in this collection is meso-level mobilities (perceived at the scale of the human senses), it also is suggestive of potential ethical and representational dilemmas that will need to be addressed in researching micro-level sub-perceptual mobilities (of electronic data, disease vectors, or neuro-chemical elements) and macro-level global mobilities (of migrants, capital, or cultural media), not to mention the potential for envisioning actor-networks as constituting more or less continuous terrains for mobile action across scales.

Many of the contributors to this volume are based in the UK, and it may be worth briefly reflecting on the specificity of British research cultures, and the ways in which mobilities research has been taken up in a number of European locations (especially in Denmark, Norway, Germany, Switzerland, and France), but has made more limited inroads in the United States. Within the European Union funding bodies have encouraged qualitative mobilities research insofar as it intersects with very concrete policy concerns such as traffic reduction or incorporation of migrants. European qualitative research is often combined with fairly sophisticated quantitative data, and in some cases the mapping and visual representation of complex data sets. In the United States funding has tended to go to existing disciplines such as Transport Studies or Urban Policy and Planning, employing traditional quantitative and 'objective' scientific methodological paradigms, or to research oriented toward more technical aspects (i.e., Mobile Communications) or managerial concerns (i.e., Tourism and Hospitality). In both cases there is little interest in affect, place, meaning, culture, or representation outside of the humanities disciplines. Nevertheless, a Pan-American Mobilities Network has just been formed in 2009, and will certainly begin to promote alternative methodologies that can bring new qualitative perspectives to public policy debates.

This is an exciting time to be engaged in mobilities research, when so much interest in the past, present and future of mobility is percolating through the fields of cultural geography, sociology and anthropology of mobilities, as well as being engaged with in the humanities, film and media studies, and various schools of theory (feminist, queer, architecture, design, etc.). As many cities, societies, and citizens face up to the limits of current forms of mobility – at the same time that many new technologies of transport, mobile communication, and real-time mapping are coming into practice – researchers are finding it imperative to move beyond existing simplistic understandings of why people move and what mobility means, and beyond simplistic techniques for representing and mapping movement. Mobile methodologies will be increasingly crucial to these proliferating fields, and the grounded dialogue between theory and methods presented in the contributions to this volume are one very necessary starting point for this new interdisciplinary undertaking.

Mimi Sheller
Drexel University
Philadelphia, USA

List of Contributors

Dr David Bissell is a Lecturer in Sociology at the Australian National University in Canberra, Australia.

Dr Katrina Brown is a cultural geographer working in the field of landscape and rural governance at the Macaulay Land Use Research Institute in Aberdeen.

Dr Nick Clarke is Lecturer in Human Geography at the University of Southampton.

Professor Sara Cohen is a Professor at the School of Music and Director of the Institute of Popular Music, University of Liverpool.

Dr Sara Delamont is Reader in Sociology at the School of Social Sciences, Cardiff University.

Professor Lise Drewes Nielsen is the leader of the research unit Space, Place, Mobility and Urban Studies at Dept. for Environmental, Social and Spatial Change at Roskilde University, Denmark.

Dr Ben Fincham is a Lecturer in Sociology at the University of Sussex.

Dr Malene Freudendal-Pedersen is a member of the Space, Place, Mobility and Urban Studies research unit in the Department for Environmental, Social and Spatial Change at Roskilde University, Denmark.

Dr Katrine Hartmann-Petersen is a member of the Space, Place, Mobility and Urban Studies research unit in the Department for Environmental, Social and Spatial Change at Roskilde University, Denmark.

Dr Eric Laurier is Senior Research Fellow at the School of Geosciences, University of Edinburgh.

Dr Brett Lashua is a Lecturer in the Carnegie Faculty of Sport and Education, Leeds Metropolitan University.

Professor Gayle Letherby is Professor of Sociology in the School of Psychosocial Sciences, University of Plymouth.

Dr Mark McGuinness is Head of Geography at Bath Spa University.

Dr Lesley Murray is a Research Fellow in the Social Science Policy and Research Centre at the University of Brighton.

Professor Mimi Sheller is Director of the Mobilities Research and Policy Center and Professor of Sociology in the Department of Culture and Communication, at Drexel University, Philadelphia, USA.

Dr Justin Spinney is a Research Fellow in the multidisciplinary research group on Lifestyles, Values and the Environment (RESOLVE) at the University of Surrey (UK).

Dr Neil Stephens is a Research Associate based at the Centre for the Economic and Social Aspects of Genomics (Cesagen), Cardiff University.

Dr Ian Walker is Lecturer in Psychology at the University of Bath.

Introduction

Ben Fincham, Mark McGuinness and Lesley Murray

It has long been recognised that the world appears different on the move – we understand it and relate to it in distinct ways from when we are still (Thrift, 1996; Cresswell, 2003; Urry, 2007). Our relationship to each other, space, time and place are mediated by our movement through the material and the social world. The observation that such relationships are influenced largely by movement, flow or mobility is by no means new. Baudelaire's 19th century flâneur understood that it was not just to be in the city that brought one close to it, it was the movement through it that allowed them to truly experience it. The act of observing whilst moving – surfacing and disappearing in the crowd – marked the flâneur as a scholar of the street, as Benjamin puts it 'botanising the asphalt' (Urry, 2007: 69). Some contemporary scholars are realising that the connection that Baudelaire offers with the mobile and the immobile – the traveller and what is being travelled through – are as important as ever, but our means to access meaningfully with our experience of a world constituted in a dizzying array of mobilities – actual, virtual, super fast, overt, covert, deadly or dislocating – have been largely ignored. In the 21st century it does not appear reasonable or indeed possible to suggest one could stroll through the world, merely observing. Particularly in 'developed' economies this luxury is compromised by an age characterised as hypermobile (Adams, 1999) for some, and hypomobile (Murray, 2009a) for others. Our obsession with speed, as Virilio points out, leaves us at the mercy of the technologies delivering increasingly rapid transfers of almost everything (Virilio, 1977). This frenetic dynamism in turn produces stark inequalities as the gap between the mobile rich and mobile poor widens. The sedate and contemplative implications of the technique of flâneurie stand in contrast to the desire to connect with and understand the mobile polarisation of the 21st century. Perhaps it is not enough to imagine that a nuanced understanding of an increasingly mobile world will be gained by simply passing through it and observing in the mode of the flâneur. Unlike the more localised world of Baudelaire's Paris, the types of mobilities connecting disparate populations and parts of the world – physically and virtually – are many and various and there is a growing feeling that we have not adequately attended to the range of techniques and their intersections that we could employ to better understand a world constituted in movement.

In geography, urban studies, sociology and elsewhere there is heightened awareness of the importance of better understandings of the movement of bodies through space. Explorations of mobility have considered commuting by car (Edensor, 2003) touring by coach (Edensor and Holloway's, 2008), walking in the city (Anderson, 2004; Hall *et al.*, 2006), driving practices (Dant, 2004; Laurier *et al.*, 2008), cycling (Fincham, 2007; Spinney, 2006) as well as methods such as go-alongs (Kusenbach, 2003), soundwalking (Lashua, *et al.*, 2006), autoethnography (Fincham, 2006; Watts, 2008) ethnomethodology (Laurier, 2005) and photo-diaries (Latham, 2003). Mobile methods are also emerging from social studies that were not previously concerned with mobile experiences but have adopted a mobile approach (Pink, 2007; Emmel & Clark, 2007). However, there has been no comprehensive text written about *how* we should best study mobilities and the potential contribution of mobile methods in enriching our understanding of the world. This collection attempts to redress a dearth of writing on approaches to studying a world constituted in mobile practices and considers methodological responses, technologies and representational strategies designed to more fully inform our understanding of people's experience of movement through space.

The methodological challenge

A sustained emphasis on movement provokes interesting methodological questions: how do we research and represent mobile experiences: of being in place momentarily, of passing through, of being 'in-between'? Can existing social scientific research methods that slow down and freeze experiences (the interview, the focus group, the survey) adequately capture mobile experiences, practices where the context of movement itself may be crucial to understanding the significance of the event to the participant, rather than being simply 'read off' from destination points and origins?

In *Mobilities,* Urry (2007: 39–42) briefly introduces aspects of the methodological challenge posed by the study of mobilities. He outlines the use of observation, the techniques and technologies required to capture these experiences, he also suggests mobile ethnography – interviewing, or being with, on the move – as part of the toolkit of the student of mobilities. Urry suggests the use of time-space diaries and the analysis of texts, 'web-sites, multi-user discussion groups, blogs, emails and list-serves' to get at the 'imaginative and virtual mobilities of people' (40). Further the use of literary or poetic techniques is highlighted as a means for eliciting 'atmosphere'. Memory work and the deployment of meaningful objects are outlined as ways of excavating 'private worlds' and senses of place in time. In a technical sense Urry writes that the tracking of objects around the globe by means of digital mapping may allow us to make sense of the cultural significance of relative worth and value through movement. He then talks of analysing hybrid systems and 'places of movement'. Of these he says 'such hybrid systems that contingently produce distinct places need examination through methods that plot, document, monitor and juxtapose places on the go' (42). He finally suggests a need to develop ways of exploring 'transfer points' or 'places of in-between-ness' (42) and the *im*mobile as part of the nexus of a mobile society.

As is suggested by the above, a narrow conception of movement as being best understood in a material, transportation sense is being dismantled. The dawn of the virtual age lends a level of connectedness and movement of information, culture or perhaps hegemony unknown in previous times. The impacts of such developments on our conceptualisation of the implication of increasing mobilities are immediate. Whilst the transfer of information via new technologies has increased, any anticipated fall in business air travel has not materialised. In fact the speeding up of movement in the virtual realm has its parallel in the physical realm. Our relationship to movement in the material world has become more frenetic and the technology designed to relieve the pressure of hypermobility has simply assisted the velocity of modern life. At the same time the antithesis of this hypermobility is the increasing disparities in access to space and resources. Social inequalities are intensified by the unevenness, the crumpling of time and distance as specific forms of mobility are available to specific populations (Massey, 1994, 2006). Whilst numerous studies have explored this unevenness (Church *et al.,* 2000; Hine & Mitchell, 2003; Kenyon, 2006; Lucas, 2004; Raje, 2004) few studies have emphasised the methods best applied for its fuller understanding.

Adopting methodologies that are distinctly mobile is a relatively new approach to research, requiring an epistemology that draws from both existing methodological paradigms and from contemporary approaches to mobilities, most notably Sheller and Urry's (2006a, 2006b) 'new mobilities paradigm'. In setting out their 'new mobilities paradigm', which places mobile practices and cultures at the centre of social processes, Sheller and Urry (2006a, 2006b) and Urry (2007) argue that the exploration of mobile bodies in mobile contexts requires a range of approaches that often diverge from more traditional methodological approaches. How-ever as this collection demonstrates, most mobile methods are rooted in existing more static methods, which have been innovated.

Organisation of contributions

The chapters fall broadly into two parts. The first section addresses questions of rationales for adopting mobile methodological approaches and the second is intended to demonstrate the ways in which researchers have employed mobile methods in various projects. Whilst the division between these two sections is not entirely straightforward contributors to the first section have often used examples of research to demonstrate their rationale and some contributors to the second section have included elements of a broader epistemological rationale – our intention is to present a justification for, followed by demonstration of, distinctly mobile methodological approaches and practices.

Interpretative contexts – The world looks different on the move

The methods discussed in this collection are predicated on research in context, within the physical and social space within which each study takes place. Everyday activities are therefore so embedded in space that to carry out data collection, for example interview in another unrelated space, can limit the potential of the data – it removes the immediate relationship between the interviewee and the emotional and social space that is being discussed. Each of the chapters attempts to embed the data gathering and to a certain extent the analytic as close to the mobile practice as possible. This closeness is an element that is variously described – can an appropriate proximity be achieved through the use of technologies or does the researcher have to be present? The commitment to mobile methodologies is very new and those writing in this collection are at the forefront of a growing realisation that in order to accurately interpret, represent and understand a world increasingly constituted in mobilities the social sciences need to develop effective techniques for studying such a world.

Attending to the everyday

The idea of the everyday, and even the mundane, is a concern for practitioners of mobile methodologies. The 'taken for grantedness' of many types of mobilities presents rich terrain for the social scientist aiming to make the familiar strange. It is the pervasiveness of mobilities that conceals it as a key factor in shaping our everyday experiences – movement is *too* familiar to be of note and as a result the mobile constitution of the world goes unexamined and evades the critical gaze. In this collection Freudendal-Pedersen, Nielsen and Hartmann-Petersen note the ways in which we have previously perceived mobilities, and the social impact of mobilities being largely through transport studies, and more specifically transport modelling. A narrow conception of mobilities as inhabiting these specific and quantifiable realms ignores the amorphous spread of movement throughout other areas of our lives. They quite rightly suggest that whilst such approaches have a place they cannot help us understand the everyday experience of living in a world constituted by mobilities. Their call for more personal and nuanced approaches – what they call critical mobility research – resonates with the theoretical orientation adopted by the various authors in this volume. An interesting engagement with the everyday experience of the impacts of mobility is demonstrated by Bissell. His interest in the relationship between experiences of stasis and movement opens the debate about our reactions to, first situations where our anticipations of movement are confounded, and also where moments of stillness or disengagement from mobility are increasingly rare. Also in this collection experiences of everyday car use (Laurier), journeys to school (Murray), cycling (Brown and Spinney) and experiencing the built environment (Lashua and Cohen) all introduce aspects of the everyday and often the mundane as being worthy of scrutiny.

'Being there' – practical suggestions for 'capture' *in situ*

The importance of 'being there' is highlighted by many of the studies presented here. The contextual and partial nature of interpretation or knowledge makes the position of the researchers important to not just the data generation process but the analytic process. In their descriptions of 'walking tours' Lashua and Cohen demonstrate the importance of being there. As with several of the contributors they uncover the importance of mundane journeys in the construction of the practice of contemporary musicianship. These are the sorts of data that are difficult to get at using more traditional, static techniques as participants often screen out the less interesting – as they view them – parts of their everyday lives. However, it is the mundane and everyday that constitutes most of our lives. Lashua and Cohen also highlight the importance of the built environment in determining the telling of our lives to others. This is a point made by Murray when discussing the effectiveness of *in situ* studies of children's journeys to school. The observations made by the young people in her study were influenced by their phenomenal experience of the journeys – what they saw and felt – and these phenomenal reactions were captured by Murray because she was there. The idea of capturing the mundane and the unanticipated is well illustrated by Clarke in his chapter on working holiday makers in Australia. Clarke went to Australia expecting to find 'frictionless nomads' and instead found 'homesick poms' [slang for British people]. The autobiographical is present in several of the contributions, and the use of these intimate first person accounts is a return, as Letherby points out in this collection, to a tradition with antecedents in the symbolic interactionist tradition of C. Wright Mills and others in the Chicago tradition. In autobiography the researcher is omnipresent and the analytic frames that support observation are applied once the initial processing of experiences has occurred. In this way the marriage of method and analysis is brought closer together with the in context reflection on experiences informing any *post hoc* analysis. Also, as Letherby effectively demonstrates, it is difficult to see how an element of the autobiographical can be absent from research. Her point is that an acknowledgement of this – particularly in mobile research – permits levels of reflection and reflexivity not appreciated from other epistemological standpoints.

Interestingly there are levels of commitment to 'being there' represented in the collection. The commitment of some insisting that they need to be there in person is somewhat challenged by those that think that similar depths of understanding of practices can be garnered through the use of technologies – particularly video.

The social constitution of mobile practices

A key theme addressed by several authors in this collection is that social relations are changed through mobile practices, however the social sciences in particular have been relatively slow in recognising the methodological implications of such as observation. The sets of methods at the disposal of researchers are characterised as being insensitive to the contexts in which social relations are constituted through mobilities. As Bissell and Murray explain the experiences that people have on the move – and in Bissell's chapter – are key to making sense of the shifting and contextual nature of social situations more widely on the one hand and people's sense of their place in such situations on the other. Through such arguments the requirement to develop a set of methods to reflect this epistemological question becomes evident. It is not appropriate to use methods that, by their nature, struggle to accommodate the contextual challenge posed by a world composed of mobile practices.

As more established methods attempt to document the social world, the authors of these chapters recognise the productive nature of mobilities. The dynamism of the productive nature of mobilities has to be reflected in the methods used to study them. Lashua and Cohen illustrate the productive dynamic through their study of music making in urban space. By examining the physical movement of musicians, music making and the relationship with 'immobile' urban forms Lashua and Cohen argue that music making is both a productive and a product of mobile urban space. The multiple meanings that can be read into a single movement, as illustrated by Delamont and Stephens, are further evidence of the productive nature of a mobile take on social interactions. The *armada* – a specific kick practised in the Brazilian 'dance, fight and game' *capoeira* is shown to be understood in different ways depending on the context within which the single movement is being performed. The observation that interpretation is context driven may not be new, but the framing of the *armada* in the context of the fight shows the extent to which both the context and the mobile practice are assumed to be understood, and such assumptions often mask complexity – particularly in the academic arena [there are clearly 'common sense' understandings held by practitioners of, for example, *capoeira*]. For Delamont and Stephens part of the social terrain to be examined is embodied and as a result their interpretation of the *armada* is as much a question of how best to study the mobile body in social contexts. Their analysis of the experience of a very specific aspect of corporeal movement is a great example of the combination of practice and analysis illuminating the complexity of seemingly straightforward phenomena. It is undoubtedly the case that the researchers' immersion in the field, and in particular Stephens' participation in *capoeira* has enabled a reading of meaning into particular 'body techniques' (Mauss, 2006). Another, but very different plea for the usefulness of embodied approaches to studying mobilities is provided by

Brown and Spinney. In their chapter on representing the experiences of cycling they provide a compelling argument for the primacy of embodied experience as not just contributing but *defining* our worlds.

Re-evaluating ethics

The ethical implications of doing mobile research present us with a set of problems and judgements that, whilst not unique to employing mobile techniques, accentuate ethical dimensions to research often easily buried or disguised by more conventional methods. In his contribution to this volume, Walker offers an important discussion on the positions that can be taken by researchers in relation to levels of sensitivity to ethical considerations. As he suggests observing the world without any recording of data has a different relationship to ethical sensitivity to a time stamped covert video recording, for example. This then begs the question to what extent can consent be granted in mobile interactions – it is simply not possible to inform a car driver that has passed by and been recorded by a researcher on a bicycle using head mounted video equipment that they have just become part of some research. This collection challenges the increasingly a-contextual formulation of ethical governance in research. The sensitivity of particular methods to the uniqueness of contexts demands sensitivity in the application of appropriate mechanisms of ethical governance.

Video and the technological fix

An interesting and exciting development in research generally has been the introduction of contemporary technologies to assist in the research process. Several of the authors in this collection describe the innovative ways in which technology has been used to inform research – particularly in an attempt to bring the data closer to the point of generation. Most clearly the development of video technology has allowed researchers to capture data in new and interesting ways. The introduction of visual data to studies has elicited particularly pertinent observations about the social world – and this is particularly true in studies of mobilities. Brown and Spinney explain that the importance of 'being there' has often been compromised in certain types of research by practical obstacles. In cycling research 'going along' with participants is often 'arduous', difficult' and 'dangerous'. For Brown and Spinney the key issue was that informants needed to be able to communicate their experiences and feelings whilst engaged in their practice – mountain biking, urban cycling or whatever it might be. Whilst acknowledging criticism of the use of video as performing the opposite of Brown and Spinney's intention, namely that it disengages and distances the researcher from the participant, they provide a clear rationale for the usefulness of the use of such technologies. In particular the use of headcams in mobilities research offers entirely new perspectives on activities where going along simply is not feasible or desirable. They also point out the value of having a technology that can record solitary pursuits whilst they are happening. Eric Laurier also uses video technology to great effect in his study of 'life in the car'. His positioning of cameras inside motor cars and subsequent analysis of the social interaction of the occupants – drivers and passengers – is made more interesting, Laurier argues, for the lack of presence of the researcher. It is also in the method that the application of established analytic tools becomes revitalised. In this case conversation analysis is used to understand the nuance not just being heard but also being seen. The benefit of video for a view from participants is offered by Murray where she gave schoolchildren cameras to record their journeys to school. In this instance the intention wasn't necessarily to capture naturalistic data in the way that Brown and Spinney did, but to allow an affective, emotive account from the students' conscious selection of elements of their everyday life. Walker used video in a different way again. Mounting a camera on his bicycle he recorded the instances of motorists overtaking him and then established their proximity to him in overtaking, but was also able to observe particular aspects of driving behaviour. With the one data gathering devise came two data sets – one more quantitative the other qualitative.

Reflexivity, situatedness and *post hoc* reflections on 'real time' data gathering

In all of the articles presented here the researchers' relationship to the research environment has involved levels of reflexivity requiring a specific acknowledgment of the mobile in not just the social context but also the research context. That is to say that in terms of data gathering and *post hoc* reflection the affect of the mobile on the researcher or the research participant is turned into data itself. In this sense the idea of real time data gathering and the analytic process become very closely related. The intention is that the data as accurately reflects the real time experience as possible and this in turn lends a real time or verité credibility to the analysis.

The conundrum faced by people researching mobile social spaces is how best to represent the dynamic and moving in a static format. As Brown and Spinney say in their chapter the distillation of life into written text loses 'the contexts that makes experience meaningful'. In terms of dissemination many of the techniques referred to in this collection offer themselves to innovative and interesting possibilities for the representation of research. The use of video and photographs are potentially fertile grounds for exploring how best to represent the 'real time' observations of mobile methods research.

The foregrounding of the mobile contexts of social situations and social relations accentuates the need for techniques of data gathering and data analysis that are sensitive to the specificity of interactions constituted through mobility. In this collection the authors have offered their ideas and opinions on how this might be best achieved. From the academic analysis of autobiographical accounts in Letherby and Clarke to the close reading of the narrative and visual scripts in Laurier's work the position of the researcher in relation to, not just the research participants, but the research context is paramount. As Brown and Spinney attempt to get as close to the 'ride along' as they can and Lashua and Cohen conduct mobile interviews to elicit particular types of real time responses to the built environment, the acknowledgement of the worth of thinking about mobilities as requiring particular techniques is clear.

Final words...

The realisation that there is something different in social relations constituted in motion, that data generated on the move is somehow distinct in character or studies that recognise that motion is of central importance to understanding certain aspects of the social world is gathering pace. In this volume studies of childhood experiences, labour markets, risk, psychosocial subjectivity and many others, there is a 'writing in' of mobilities as key to orienting new and innovative ways of interrogating the social world.

Part I
Driving the Mobile

1
Contextualising and Mobilising Research

Lesley Murray

Most of the chapters in this collection are either predicated on, or make reference to, the value of mobile methods in researching in place or 'being there'. This is a critical element of mobile methodologies that presuppose the centrality of mobility in everyday life and societal structures (Urry, 2000, 2007; Cresswell, 2006). This methodological approach, along with methods that best capture mobilities, are considered crucial in enhancing our knowledge of the world. An exploration of the notion of researching in place, in spatial context, can tell us about both these elements of mobile research. Firstly, being in place is always relative to another place. A specific spatial context can only be given meaning in relation to another, or to a space it is not. These spatial contexts are then linked by mobility, be it corporeal, visual, audial, imagined or virtual. Secondly, researching in situ requires consideration of the research methods that make this possible. It can both demonstrate the need for mobile methods as well as their application. This chapter is based on my research, which explored the mobilities of the school journey and involved consideration of the same research questions with a group of young people in different spatial contexts. In doing so this chapter describes the process, outcome and implications of research that is mobile, and involves a number of different spatial contexts. It addresses the question of what happens when we research the same question in different places.

My research explored the co-construction of risk, mobility, motherhood and childhood in determining mothers' and children's mobilities. Although the methods adopted with mothers were relatively static, I used both visual and mobile methods in researching with young people (see Murray, 2009a, 2009b). Twenty five young people from a range of social backgrounds and in different urban contexts filmed their journey to or from school, often describing their feelings and responses to mobile space as they travelled. Videoing was followed by film-elicitation interviews, with the young people's footage acting as a focus of discussion. The young people thereby described experiences and feelings about various aspects of their journey: the chosen route; risks encountered along the way; who they travelled with; how decisions are made about how they travel; and changes they would like to make, in at least two different contexts: the mobile context of their journey and the relatively static context of their home. Three themes emerged from this methodological approach: the making of spatial contexts through mobility; the distinction between responsive and contemplative contexts and the contextualised production of knowledge; and the overall enrichment of data through multi-context and therefore mobile research.

The importance of context – making space and place

> Jimmy: We've got about four ways to school. Through the Rec., straight down, up here left or right... This way seems to be quickest and we've got that secret path. Well it's not very secret but that's cool.
> Jimmy's friend: Joe copied us.
> Jimmy: Yeah, their route's exactly the same as ours now.
> (Jimmy and friend, video)

For a number of young people involved in the journey to school research, it was very important that they decided on their route to school and that they retained ownership of it. They constructed certain spaces as part of *their* mobility. They have carved out a path with various markers that symbolise this ownership, a process that is considered by Georg Simmel (in Urry, 2007) to be, at its core, a distinguishing feature of human civilisation. In the videos they point out their chosen landmarks, such as houses they consider to be particularly attractive, along their chosen route of the day.

> There's the flats...There's the park, more flats, Forestside. There's lots of dogs, people walking...We're going to go over the park... There's the police station, the bus stop...There's the big queue of traffic...That's a nice house...There's the zebra crossing. There's a recycling truck. There's a café – people go there to eat breakfast. There's the co-op. Some nice hanging baskets. There's the big flats. There's the greengrocers over there...There's the new shop... There's

the library and here we are at our school. That's our journey. (Lucy, video)

That's the big house at the end of the road then you've got to cross the main road which is absolutely horrible. I like running across it. It gets warmer up here than down there. One of my friends lives in one of these houses...This is one of my favourite places to walk up. I like it 'cause its curvy...This is my favourite place to walk up here...This road is tiring to get up but it's a really nice road ...This is the road we have to cross as you can see it gets really busy to cross but when we get across it its my favourite favourite part of the journey... Here's another house I like and then we come to another house I really like that has a pond on it. We're really close to our journey ending. That's my house, I like the pond house. I bugseyed it. I bugseyed it the first time I saw it. It's a really pretty house. (Lily, video)

In this way the young people are producing their mobile space, a space that would not emerge from a static interview but is the product of the mobile methods adopted. As discussed by Brown and Spinney (Chapter 9) and Pink (2007) the production of space in this way becomes associated with the research process and the filming of their journeys. As they made their journeys within the research context, the young participants were able to ascribe meaning to space and to attach certain emotions to particular spaces.

I hate the main road. We're so scared of it... Now we have to dodge the cars and that's not pretty. (Jasmine, video)

Jasmine and Lily explain that they pass a certain house every day and smile at the woman who lives there and she smiles back. They not only have marked the space but have initiated a social contact, perhaps to enhance their personal security. Throughout their journey they demonstrate extremes of emotion along the way, from really stressed when they cross a busy road or encounter some older children from their school, to enjoyment of the houses and people they pass.

You can turn off here to go to school. No, we'll go this way. It's very peaceful. I like walking through the park better than the roads because you get to see all the...in the morning all the dew on the grass and stuff. We're going past the tennis courts and we go along that lane and then we get to the school. (Molly, video)

Certain spaces evoke particular emotions. For example, green spaces seem to have the therapeutic quality that Conradson (2005) describes in terms of the emotionality of space, representing tranquillity and relaxation, whilst also representing a potential for challenge and excitement. The video data provided an insight into these emotional responses to space.

Responding and reflecting

My study also used the visual method of film-elicitation with children and young people, using a film in a semi-structured interview to provoke discussion and re-situate them in the space (see Banks, 2001 and Bryman, 2004 for examples of the similar method, photo-elicitation). Having filmed their journeys, the focus of research with the young people then moved to another context that was in some respects as mobile in that it involved them watching the video of their journey, a mobile representation of it that allowed a step by step re-making of their mobile experience. The young people could then reflect on their representations of their journey by watching the video recording. The use of visual methods in complementing mobile methods particularly facilitates this reflection (see Dant, 2004; Holliday, 2000, for examples of video methods). Often the process of reflection not only involved the young people, but their families and the researcher, beginning the cycle of making and re-making the journey through ascribing new sets of meanings as the video is audienced (Rose, 2007) in a number of contexts.

As discussed, different mobile spaces such as outdoor walking spaces and insulated spaces of buses and cars can evoke different responses. As Urry (2000) argues, the culture of 'dwelling-in-the-car' results in insulation from the outside world, in an externalisation of risks. This is borne out in videos produced by car passengers. The young people videoed mainly their view of the inside of the car as the outside world seemed beyond their visualisation and their senses related only to the inside space. The car, therefore, can produce a different spatial context within a broader space and in doing so reduces the scale of dwelling. The extent of this shrinking spatial awareness, as it relates to the car but also to other modes, can therefore be examined through mobile research that considers the sensory responses to different spaces. As Lucy (interview) said 'you could see all the traffic and not much of the countryside'. During Lewis's video it was clear that this is particularly important for children due to their height, as his view was significantly restricted by the level of the windows. Although Lewis appears to be filming the outside space, as he is pointing the camera in this direction, the image is blurred and it is indeed this distortion of outside space that is the product of the semi-private space of the car that he is filming. Similarly, Ben gives particular responses in the socio-spatial context of his car journey with his mother, which reflect that particular context.

> Might do this video camera more often cause it seems to stop my mum's road rage…We don't normally get in the car. I normally get the bus or walk or cycle cause it's easier. We always end up slacking in the morning if we go by car…I prefer cycling personally. Just making the exception in the car cause I've done in my knee…I'm just filming the car in front. That's not very exciting. I'll film the dashboard instead. (Ben, video)

Ben appears fairly carefree and relaxed during his journey in the car, even though he claims to prefer walking to school and is only travelling by car because he injured his leg. In his interview, however, he indicates that he is more cautious about his journey to school, discussing a number of specific risks as well as broader discourses of risk.

> If you're running late there's not much you can do except run which isn't exactly very enjoyable, especially if you have a school bag on…you can get frightened [in the dark] 'cause of course someone could jump out of the bushes…Well, if you've had trouble the previous day at school you could be worried about the person…seeing the person outside school before school. Apart from that it's a pretty safe journey. (Ben, interview)

> …there's one when a man…Crimewatch funnily enough…and a girl was being followed and a man grabbed her and she ran across the road to the nearest adult shouting 'mum' and the adult assessed the situation and said OK. That keeps me reassured. When I first went to school my mum would say if this happens blah blah blah. (Ben, interview)

The video data and interview data therefore did not necessarily produce parallel accounts from participants, especially in exploring subjective notions such as risk. However, in line with Denzin and Lincoln (1994) the process was not one of seeking truths and so this inconsistency was not interpreted negatively. Instead it demonstrates one of the strengths of the mixed methods adopted here in allowing the emotionality of the journey to be explored alongside more considered responses, responses based on reflection and in different spatial and socio-cultural contexts. There could therefore be no triangulation or validation of data but a rather a new set of interpretations and analysis. The different contexts provided different layers of data to be made sense of.

For a significant number of children, it was evident that issues which are relevant when they are emotionally engaged with their mobility are less relevant when they are disengaged and static in film-elicitation interviews afterwards. A number of children mentioned fears they had, particularly about being on their own, during their journeys but did not mention these when asked in the interviews about any worries they might have about the journey. So even though they had just watched their journey, they gave different responses when out of the spatial context of the journey. Molly (video) talks of 'get[ting] a bit scared sometimes, whereas Evie (video) said she was a 'bit nervous of the park 'cause there might be like strange men'. However, later in her interview, she says that she has no worries. Both Evie and Molly travel through the same park, although on separate journeys, and it could be that there are often strangers in the park or that they encounter the same strangers. An important aspect of Evie and Molly's risk therefore, even though they travel separately is that it is shared or collective (Tulloch & Lupton, 2003). It is a risk that is experienced in response to wider discourses, both local and global (Holloway & Valentine, 2000) that results in their experience of the same space in a similar way. For example, they could be related to prevailing notions of the riskiness of 'strange men' (Valentine, 2004).

Film elicitation not only provides a focus of discussion but contextualises it as participants are watching their experiences of a particular place. However participants are no longer immersed in the multi-sensory experience of the journey, such as the noise from busy roads, the sense of tranquillity of the parks or the smell of petrol and grass cuttings. They are experiencing it only in relation to the two dimensional images on the screen and the limited quality sound recordings of the video camera. The experience is a mobile one, but their range of emotional responses to space and mobility is reduced. Nevertheless this altered representation of their journey provides a context for reflection about the journey. The combination of film and film-elicitation, therefore, produces a new set of emotional responses to the journey as well as giving some insight into more direct emotional and sensory responses. This leads to insights and understandings that are particular to these methods (Banks, 2001).

In this study, the film-elicitation interviews took place in participants' homes, with the video tapes of their journeys played for them on their televisions.[1] The method of film-elicitation interviews provided freedom and flexibility in that young people could choose whether they answered questions during the showing of their film or afterwards. The interviews provided an opportunity to follow up on some of the issues raised during filming and were an opportunity for children to address issues in whatever depth they wished. The reflective context enabled participants to recall events that they had not mentioned during their videos. Some of these events, such as Lily and Jasmine's incident with a BB gun, appeared to be quite critical in terms of young people's risk and mobility yet were not mentioned during filming. The following quotes are related to the same location in their journey, one from the video and one from the interview.

> That's the big house at the end of the road then you've got to cross the main road which is absolutely horrible. I like running across it. It gets warmer up here than down there. One of my friends lives in one of these houses. (Lily, video)

> Somebody threatened us with a gun, a BB gun, to shoot us. We phoned the police. It was near [our friend's] house…We stayed near Jade's house and phoned the police from a mobile. The police came and me and my friend walked down further and the police came and they were jumping fences and everything. But they never caught them but they have now and they've put them on a list so that if they do another thing wrong…They go to our school… When we went back to school I had pictures of it shooting 'cause they put the gun to our head. I had a picture of the gun. We walked to school on the Monday. (Lily, young person, interview)

It is surprising that Lily did not include this incident in the video commentary of her journey, particularly as she did include reference to the 'friends house', which is now particularly associated with the incident. Lily's relationship with this space is represented in yet another way after further reflection on the journey, during a follow up email two years after the original fieldwork. Here Lily recalls that the incident occurred after the research, perhaps because it was not represented in the video, which had just been circulated in edited form to participants.

> A few months after we made the video these boys were messing around and pointed a fake gun at my sisters head and were so frightened for a week we walked with my mum. (Lily, two years after original research in email)

As well as placing the incident in another timeframe, Lily's reaction seems stronger and she now says that they 'were so frightened that they walked with their mother', having suggested in the film-elicitation interview that they had carried on as normal after the incident. The temporality of research methods, therefore, is a significant factor in influencing reflection and re-situating experiences. Despite this particular incident not being mentioned in their video, it was apparent during their journey that both Lily and Jasmine were fearful of other children, both appearing agitated during their filming when older children walked near them. Again their responses are different in each research context.

> There's more people from our school. Oh God should we walk on the other side...We go straight up this path and there's some chavs but I'd better not say or they'll beat me up...(Jasmine, video)

> When I walk past them they keep on swearing and they kept on swearing at me and drinking beer and smoking and there was one that when they were smoking they started to spit at us...Because we get scared of teenagers...I'm a bit...but once I know one...there these teenager that go pow pow to me but I hit them and it's so cool so now I like it. They're scared of me. (Jasmine, interview)

Some risks, therefore, become more critical with time as other sites of risk such as roads are more critical as they are directly experienced. It is possible that prevailing discourses of risk (see for example Furedi, 2008) may amplify particular risk experiences in different spatial and temporal contexts.

> Up top there that's the bit I hate. Dad says be careful on that road cause some people do normally die on that road. (Jasmine, video)

The experiential element of walking and talking in this research therefore appears to present little opportunity for reflection but is nonetheless indicative of the 'doing' of everyday life, which is often on the move. However, it appears also that sometimes there is a need to immobilise, to enable reflection, as responses to stimuli of everyday life are not only based on immediate surroundings but on personal experiences throughout the lifecourse and on influential discourses produced through socio-cultural structures and processes.

> Not all my friends are allowed to walk home on their own...Some people are allowed out more than me and some aren't. I'm in the middle (Joe, interview)

> My friend Phoebe is allowed to go on her own but her mum can see her from the kitchen window playing in the park...Some Year 6s and Year 5s who live near the school walk. (Lucy, interview)

In addition, the young people were more likely to discuss issues not directly related to the journey during the interviews afterwards and these issues tended to be more global in scale. For example, the young people were more likely to discuss broader issues of freedom rather than specific risks in relation to their independent mobility. Again, this could be related to both spatial and temporal contexts of the research.

Enriching data: Contextualising ambivalence and mobility choices

The use of different methods and in different contexts, therefore, provides insights into the everyday mobility of young people that would not have been possible using static methods. Another example of the enrichment of data relates to mobility choices around car use. Here, the ambivalence surrounding travel by car is captured through different responses in different spatial and temporal contexts.

> I don't like cars. Well...I like cars but I don't like going to school in cars. (Jimmy, interview)

> I prefer going by car 'cause it's not as cold and you don't have to wait as long and just because it's more comfy and you're in your own car and stuff. It's warmer cause the bus is quite cold and it leaks and stuff. ...I'd probably say [I like] the bus [the best]...cause I like to see my friends...(Megan, interview)

Megan's ambivalence may have been heightened as she was watching her journey video, which showed a bus interior with misted up windows, adding to the feeling of being cocooned (Urry, 2007). Megan talks about alternative ways of travelling in her video:

> There's [the] Park. You have to walk across the whole of [the] Park if you walk. I did once at the start of school but I was so tired when I got to school – never again. When I was walking really slowly on the way back from school it look 3 hours. We were just rolling around in [the] Park. (Megan, video)

Of course a certain spatial context is also imbued with cultural meanings. For George, during his journey to school, depending on who was present, he indicated that he was risk averse and then risk seeking.

> That's where the allotments are. I don't like them. This bus is going too fast. I think the speed limit on this road is too fast 'cause horses come along...I like it when the lights are green and we go faster. (George, video)

> Other kids. I had trouble on the first day when I got on with all my friends 'cause there were some Year 11s...The top deck is sort of split at the front is all the Year 7s and the back is the Year 11s and the Year 8s and 9s stay downstairs. (George, interview)

Although there is an element of performance of bravado for the camera, as soon as a certain boy gets on the bus, George's demeanour changes markedly and he becomes more reserved, choosing to stand and film from the front of the bus silently. It is also evident that George and his friends are excluded from the top deck of the bus due to this threatening behaviour. When another pupil gets on the bus and asks George why he is not filming on the top deck, George says he 'can't be bothered'. However earlier he had said;

> This is the part that if I'd got the school bus over 200 kids would have got on. Looks like the teachers didn't miss the bus. I always sit downstairs, never up. (George, video)

> These boys are Year 11s and 10s. They also think they're hard. Let's see what happens when they see the camera. (George, video)

George's risky experiences on the bus are not his only critical experiences. When he was younger George had a risky experience on a car journey with his father.

> When I was little I fell out of the car when it was on the motorway. My dad was driving. I didn't have my seat belt on and I went to get something and opened the door. The car behind saw me and skidded to a halt and luckily I managed to get out without a broken bone. It was in the early hours of the morning. It was when I was about seven. (George, young person)

This is a serious incident and obviously affected him. However, it did not emerge when asked about past risky experiences, but rather almost 'by accident' when he was talking more generally about going in the car with his father. More current experiences, such as bullying on the school bus, were recalled instantly, whereas this was embedded at a deeper level, but was nevertheless as significant an issue in his overall mobility history.[2] When questioned about this incident, it emerged that this accounts for George's worries about cars and his preference for travelling on the bus, even though he is currently experiencing risky events on this form of transport. The different contexts of the research allowed a filling of gaps in George's story, providing an enrichment of the research data that may have been denied using more static methods of research.

Conclusion

There is a range of contexts that impact on the outcomes of research – both socio-spatial and temporal. In addition, the context created by the research can produce mobilities. Young people produced their mobility through engagement with the research and changing their journeys to school for the research process. The research described here has considered the different outputs that are produced from carrying out research in different spatial contexts. Similar results, which will be published at a later date, have been found in teaching exercises based on asking the same research questions in different contexts, carried out at the University of Brighton[3] and Bath Spa University.[4] Rather than using the different results in a triangulation process that seeks validation, the different sets of data were considered to enrich the overall data. Mobile research is therefore not about finding 'truth' but about investigating placed narratives. In addition, as the research demonstrates, such mixed method mobile research is more likely to capture the taken-for-granted, mundane elements of everyday life. Mobile research can enhance knowledge about how we 'do' and how we experience what we do in different contexts. At the same time, such research demonstrates how mobility is productive of space and place, and how space and place are productive of mobilities (Cresswell, 2006).

Mobile methods allow the exploration of the emotionality of everyday mobile life while it is taking place. The researcher becomes immersed in the mobile experience along with the participants, a process that is both legitimising and productive in allowing an appreciation of the more intricate sociality and emotionality of the journey. It could be argued that all research is in place and is therefore mobile. When we interview participants in their homes, the discussion will reflect participants' view and emotionality associated with their home. For broader insights into everyday life research should be mobile, both temporally and spatially.

Notes

1   With the exception of two where technical difficulties prevented this and films were watched back on the video recorder screen.
2   The notion of 'mobility history' is based on the influence of individual and collective lifecourse experiences and socio-cultural contexts of mobility on individual and collective outlooks on mobility.
3   By Ben Fincham and Lesley Murray.
4   By Mark McGuinness.

2
Mixing Methods in the Search for Mobile Complexity

Malene Freudendal-Pedersen, Katrine Hartmann-Petersen and Lise Drewes Nielsen

The world is moving. People are on the move. To many people automobility is the cement of everyday life, and without the car, everyday life cannot be coherent (Freudendal-Pedersen, 2009; Freudendal-Pedersen & Hartmann-Petersen, 2006; Drewes Nielsen, 2005). Mobility and modernity are merged (Urry, 2000; Urry, 2007; Canzler *et al.*, 2008). Late modern changes in everyday life spheres are complex, with our understanding of these transformations in a constant state of flux (Bauman, 2000). As this affects the way we perceive and analyse the world around us, it calls for new methodological approaches and new ways of collecting empirical data. Also important here is the role we, as researchers, play in the research process between research design, data collection and analysis.

The aim of this chapter is to show how mixing qualitative methods such as interviews, focus groups and interactive workshops can open up new ways of understanding the ambivalent role of mobility in everyday life. Our approach to qualitative methods challenges the transport and mobility research in two ways. Firstly, there is no qualitative research tradition in transport research and as such this analytical approach applies methods and knowledge from classical sociology to transport research. Secondly, within mobility research the aim is to show the interconnections between mobility and modernity. This is primarily done with an absence of lived experiences – the voices of everyday life. Focus has been on developing a new concept to embrace and develop a new understanding of late modern lives in the age of mobility. As such we can claim that mobility research aims at changing the world or our understanding of the world but not through a dialogue with the lived experiences. What is essential for us in our work is to get life utterances into mobility research and analysis, through mixing different kinds of qualitative methods, and especially through interactive workshops.

Mobility research requires both theoretical and methodological considerations (Hannam *et al.*, 2006; Kaufmann, 2002), as well as a discussion of the philosophy of science. This article raises some of the questions relating to how qualitative empirical research can contribute to the further development of mobility studies, both theoretically and conceptually. Our focus is on the tacit knowledge and irrationalities concerning mobility issues in everyday life, which require a rethinking of research methods. In particular, using different methods to understand the wishes, dreams and aspirations of everyday life creates opportunities to understand some of the driving forces and the taken-for-granted knowledge behind late modern mobilities. Within this critical approach, transparency of the research process is essential.

Transparency here means to create open and visible criteria in the research design processes, as this is essential for estimating the quality of the research (Silverman, 2005; Flick, 2006). In striving for transparency the following questions emerge: Why are we making specific research choices? How are we doing it? What do we de-select? Questions like these become important in explicitly considering the research subject. Therefore, in this chapter, where the research ambition is to build the qualitative research process upon the values of *openness* and *transparency*, we will explicitly present the key ideas upon which the research is based to facilitate full scrutiny of the process. We will present the ontological, epistemological and methodological choices associated with the use of in-depth interviews, focus groups and action research inspired workshops in exploring our mobile everyday world.

Choices of reflexivity

Numerous choices must be made throughout the research process, and each of them has influence on the final outcome of the analysis. These choices and opportunities can be described as reflexivity. A reflexive method (Alvesson & Sköldberg, 2000; Drewes Nielsen, 2001) implies that we, as researchers, assess and explicate a number of choices made in relation to us and our researched problem (problem formulation) and field of research. These choices relate to: Why exactly do we study this problem? Which theories will be used? Which methods will be used? How will we orchestrate the analysis? Who will assess our results? Who will use our results? Should our results aim at change? This reflexive approach, first and foremost, requires us to be transparent and explicit, and this is essential if the research produced should be open for util-isation by others. In addition, the clear visibility and rationale for certain courses of action will have a direct impact on the clarity of the research conclusions.

In order to make reflexive choices, we work with a number of pre-conceptions. We meet our research problem with knowledge attained from our education, our previous research, from diverse literature, our engagement in society, from the news etc. These preconceptions con-stitute the basis for the generation of knowledge in the research process. While they may be theoretical or empirical, the process of identifying and presenting them is essential, as they form the foundation of our research.

Methods and strategies for analysis

The following examples used in this chapter are based on a sociological approach to mobility research, which is empirically based on the move-ment of goods, people and information as well as related technologies, institutions and systems which all promote mobility (Hartmann-Petersen *et al.*, 2007). Mobility research uses a theoretical and methodological approach to understanding the spatial, psychological and social dynamics of mobility. This helps to create a better understanding of the importance of mobility in terms of behavioural change and regulation of transport behaviour, road safety and traffic demands. Earlier research in physical mobility (transport research) has, so far, focused on technology and modelling. The sociological approach to mobility research therefore requires new concepts and methods in order to understand other aspects of corporeal mobility as well as wider aspects of mobility (Urry, 2007). This poses new challenges for research in this field and, in particular, calls for transparency within the research process.

The basis of our analyses of aspects of the mobile world is, therefore, an explicit consideration of the preconception of the projects, that is to say, the ontological, epistemological and methodological assumptions which underlie the research. We discuss how these preconceptions and con-siderations set the context for the research process, the analysis, and the subsequent knowledge produced.

Ontology: The examples used in this article, are based on an ontological position that lies between constructivism and critical realism. This onto-logy is based on knowledge of lived reality where knowledge is under-stood as constructed; a construction based on the materialities of everyday life whilst simultaneously redefining the significance of these various materialities. Defining this ontological position means that we, as a starting point, have an understanding of the pressures of late modern everyday life These pressures are predicated on the choices that individuals are forced to make in comparison with past societies, in which individuals acted according to tradition (Giddens, 1991). Thus, the individual is often very much aware of the temporary nature of knowledge. This constant review and debate of knowledge creates an increased sense of risks (Beck, 1992).

In our reflexive time-pressured everyday life, many choices have to be made, from the type of washing powder we use, to whether to worry that the ice caps are melting. We make these choices on the basis of expert knowledge and form opinions based on the confidence we have in the experts we find the most sympathetic and trustworthy (Giddens, 1991; Beck, 1992). In this research, we aim to create knowledge of the interaction between the individual and the society, between actors and structures.

The starting point is a critical research aiming at *understanding* and *changing* (Thomsen *et al.*, 2005); a critique that constitutes the starting point for a transformation-oriented research, aiming at breaking with taken-for-granted knowledge, existing rationalities and preconceived ideas about the world we are moving in. Despite the use of individual interviews aiming at an in-depth understanding of individuals' perceptions of mobilities and everyday life, the critique is not directed towards an individual level, but rather aimed at the way everyday life is structured and composed through the community and societal level. Individuals master everyday life in the way it makes the most sense for themselves and those who are closest to them. The increasingly individualisation of all societal issues and challenges is a huge burden in managing everyday life. Thus the critique stems from a desire to formulate strategies for action and change through communities inter lacing with individual experience of everyday life. Communities can take various forms, as concrete or abstract, and affect lives at different levels (Freudendal-Pedersen & Hartmann-Petersen, 2006). Understanding the importance of communities in relation to individual's ontological security whilst maintaining a community perspective is essential in exploring mobilities (Bauman, 2000, 2001). Late modern individuals have difficulty in maintaining a meaningful and busy everyday life, while simultaneously formulating new scenarios for their conduct. The community level, through which some of everyday life discussions and decisions can be made, can displace some responsibility from the individual and provide the preconditions for societal change.

Conceptualising different phenomena in late modern life is fruitful in the attempt to understand how the individual values and plans everyday life. However, the changes and breaks in daily routines and actions are not clearly apparent. The objective of transformation-oriented planning can thus formulate concrete strategies for change, from an understanding of the core phenomena and their context.

Epistemology: The *epistemological* starting point is based on the types of knowledge developed within the ontological framework. Within sociological mobility research, it is of epistemological significance how our knowledge of communities, time, space, resources and movement in everyday life can be viewed as phenomena. This also implies a reflection on how knowledge can be produced in the interaction between the researcher and subject within the field of research. This interaction emerges when it is not a fixed and defined subject, but instead a process which evolves as we, as researchers, articulate it. For example, for most people, everyday mobilities are routinised and determined to a degree that is rarely articulated or subjected to reflection. Cars, buses, trains, cycling and walking are all routines of everyday life and consequently, changes in and ways of mobility are seldom reflected upon. When these routines are explored and merged with other conditions of everyday life, such as time pressure and communities, a dialogue emerges between the researcher and field of research which articulates new elements of everyday life. This means that it is in this dialogue that new knowledge of social phenomena is developed and continuously evolving, when the constructivist understanding is based on subject realisation (here the lived, mobile, everyday life) which changes from time to time. This does not, however, mean that all knowledge is non-reproducible. Rather, it is highly likely that another researcher will be able to identify some of the same phenomena at a later date or that the same method could be used in order to replicate the study elsewhere. What is important here, however, is that we are paying attention to the fact that social phenomena are not static but are rather continuously developing in the mutual influences between the researcher and the field of research.

Epistemologically this means that by following this path we achieve a particular type of information. This research does not aim to find the 'objective truth' of reality when this, from our point of view, does not exist. On the contrary, there is a danger of ending up in a relativistic trap where all the respondents' statements can be used for everything. Another danger is to subscribe to the problem of 'anecdotalism', where qualitative data is understood as a single case with a few well-chosen examples, and not as a result of critical investigation of all the data produced (Silverman, 2005). The purpose of the sociological approach to mobility is, in contrast to this, to create new knowledge and find different ways to explore, understand and change. With a constructivist aim, it is important to note that knowledge and truth are created between actors and their surrounding structures in a specific situation. This knowledge and truth is not a mirror of an objective reality. Not neglecting the fact that there is dissemination and knowledge processing between the researcher and the selected respondents, the research interest is to focus on the knowledge and conception of reality the respondents have about their daily lives, their mobility and the correlation between the two.

Methodology: The *methodological* approach is a product of the ontological and epistemological frameworks. The remaining element of the research design is to establish how we can generate data and knowledge on the research questions we wish to address. Thereby, the research questions, and the understanding of the surrounding world these questions are placed within, are what determine the methodological frame. At this point we need to ask whether we need quantitative or qualitative methods in order to find answers to the questions we ask or whether we need different approaches which complement each other in the empirical evaluation of the problem. Data production must be designed in order to ensure that it answers the research questions asked. Examining phenomena in daily lives with the focus of how they are discussed and negotiated, and how they have come to be taken for granted knowledge requires a methodological approach based on mixed methods. Different ways of creating qualitative knowledge provide the best opportunities to reveal an in-depth understanding of everyday life.

As the sociological approach to mobility research is a relatively new field of research with limited analytical- and empirical-based knowledge and research projects, the potential for complexity is high and the use of open concepts and methods are important. Therefore, a methodological approach where the empirical and the theoretical work together, an *abductive* approach, is useful here. This method is characterised by empirical work which is inspired by the theoretical, while the empirical simultaneously calls for new theoretical clarifications. Through the abductive process the two elements inform and interact with each other, and it is thus possible to achieve an empirically based and theoretically informed analytical production of knowledge (Blaikie, 1993; Alvesson & Sköldberg, 2000). In this research process, there is not a theoretical basis to be tested which determines the production of empirical results (induction), and there is not an empirical basis calling on certain theoretical frameworks for developing the analysis (deduction). Therefore, there is an abductive interaction, where the theory and the empirical data are living side by side in the research.

Our research experiences with focus on social perspectives on mobility (Thomsen *et al.*, 2005) have always been based in the empirical. This does not mean, however, that the theoretical foundations and ideas have less weight or influence on the research findings: What has been the driving force are the empirical social issues from which the majority of research project arises, and the analyses of the empirical (mainly qualitative) data are constantly discussed and developed through, and in line with, the current theoretical flows. This means that the research findings challenge the field of research by aiming to both *understand* and *change*. Where the intention of understanding relates to the traditional ideals of knowledge production, we are also using the theoretical and the empirical in order to promote new understandings with the intention to change. Thus, the point of departure has a normative basis of desire to remain critical to the reality explored from the outset where research has the ability to contribute to changes. In critical mobility research, the orientation towards change is a desire to reinstall research as an aid helping to change the most urgent social problems. In relation to mobility this is the explosive growth in transport and the growth of environmental pollution caused by transport. Critical research can contribute by facilitating the development of new processes of change and planning horizons, which provide new scenarios for future sustainable development (Thomsen *et al.*, 2005). By allowing the theoretical and the empirical to equally inform and support each other, we can contribute to targeting research findings towards changes in physical, social and cultural planning.

The methodological approach has an impact on the generation of knowledge on mobility, and the importance modernity has played in the mobile world. Identification of the factors which influence these constructions is therefore crucial. Qualitative methods allow us to explore beyond the taken-for-granted knowledge (Freudendal-Pedersen, 2009; Hartmann-Petersen *et al.*, 2007). Qualitative methods offer a way of understanding both social phenomena and the context in which they are rooted. They offer the ability to identify the construction of meaning and significance of various aspects of transport and mobility.

These methods open up the opportunity to follow new, unexpected tracks in the interview situation, focus group or workshop. If a respondent raises new or unexpected points, it is possible to follow up on these topics and explore new approaches not previously planned or foreseen. Often, this unexpected path proves to be of great importance for the understanding of the construction of meaning and significance in both everyday life and work life. Methodologically, the interaction between theoretical perspectives and empirical evidence are weighed in the research process is also significant (Freudendal-Pedersen *et al.*, 2002; Freudendal-Pedersen, 2007; Drewes Nielsen, 2005). The establishment of an interpretative paradigm or framework – understood as the clarification of the researcher's epistemological, ontological and methodological premise – constitutes the basic set of beliefs that guide action in qualitative research and is thereby a cornerstone in raising the standard of qualitative work (Denzin and Lincoln, 2000).

Case 1: Mobility in the late modern community – using interviews and focus groups

This case describes how data collected about ecology, mobility and everyday life through interviews and focus groups can be used to explore how individuals narrate their everyday life-related choices, and the unintended consequence this has to their mobility patterns (Freudendal Pedersen *et al.*, 2002). The way in which individuals reason out their everyday life decisions relating to their chosen means of transport and their mobility behaviour reveals knowledge about the condition of their hectic daily lives. The choices they make have a range of environmental and social consequences that the individual may be more or less aware of and reflective about.

The following empirical work has been gathered in the ecological village Dyssekilde, which is located 50 km from Copenhagen. The information was gathered through in-depth, qualitative, individual interviews and focus group interviews. The case was chosen from a preconception that the residents of Dyssekilde – because of their conscious choice to live in an eco-village – were aware of the environmental consequences of late modern lives and actions. The empirical basis was individual interviews with representatives from six families (nine persons in total), and two focus groups with the same nine people. Respondents were selected from as wide variation criteria as possible in relation to gender, children and work place etc.

The interviews were carried out in the form of semi-structured every-day life interviews (Kvale, 1996). Each interview took around two hours to complete and, in various ways, examined the respondent's relation-ship to everyday life, their conception of the environment, mobility etc. When creating the interview guide, the open character of the qual-itative approach provided an opportunity for empirical and theoretical preconceptions to be included in the organisation of the interviews in advance. A carefully designed interview guide can, in addition to the questions, also include the reflections of, and assumptions about, what is expected to come out of the questions, and hence the pre-conceptions underlying the different issues discussed. This clarification and recollection of hypotheses and preconceptions during the inter-views constantly reminds us, as researchers, not to follow a road validating our own preconceptions but to follow and discover new trails.

Such preconceptions are also reflected in our selection of the empir-ical group. When choosing a group of people with a specific level of education and who are living in an ecological village, we have cer-tain expectations as to what their answers might be. In order to avoid reproducing the knowledge originating from our preconceptions, it is important to be aware of what they are, in order to ensure such assumptions are avoided. The individual interviews were followed by two focus groups which contained the same people. The first focus group circled around the conceptions which the respondents had of a 'good' everyday life and the significance of mobility in this life. The second focus group was held five months later and initially dis-cussed the outcome of the first focus group. Following this the con-versation focused on concrete ideas of solving mobility-related problems, articulated by the group itself.

Discussing elements from previous interviews with respondents can further give meaning to the research when the condensed and ana-lysed knowledge is recognised by the people who actually articulated it. The timeframe between the individual interviews and the first and second focus groups allowed us to begin the analysis, and focus the design through selection and de-selection of themes. It also gave us the possibility and opportunity (after asking permission from the inter-viewees) to use quotations from interviews in the subsequent focus groups.

The individual interviews contributed to an in-depth insight into the stories each family has about their daily lives. The focus groups provided an insight into how the 'truths' individuals build their lives around are linked to practice, and the degree to which these 'truths' are negotiable or fixed in daily life. These aspects do not come out in the individual interviews since the interviewees are not confronted or opposed in their presentation of everyday life habits, contradictions and considerations. Where the individual interviews were intended to identify the problems from the individual's point of view, the intention of the focus group was to discuss and clarify these issues. This also provides an insight into how the different issues were articulated and negotiated in interaction with other individuals, which is the general difference between individual interviews and focus group interviews (Halkier, 2008). The empirical is converted into interpretation and discussion through two steps of analysis: A vertical and a horizontal analysis. The two steps of analysis each have their own function in relation to the final interpretations.

Vertical Analysis: In the vertical approach, focus is mainly on the statements of the respondents in their own context. It is the voice of the respondents which is central. The basic elements of the vertical analysis are the family stories, each of which paints a picture of the conditions in which respondents prioritise their beliefs and actions. The aim is to draw a substantial and nuanced picture of each family's life situation and to lay out the preliminary subjective perceptions of the researchers as explicitly as possible. In order to structure the families' narratives, the family stories are structured through points of reference repeated in each family. The points of reference emanate from both pre-defined preconceptions and the thematic issues derived from the interview material when respondents describe everyday life. At the end of each family story, a short summary describes our understanding of the respondents, i.e. how we perceived the situation we entered when visiting the interviewees in their homes. By making these circumstances explicit, we, as researchers, become aware of the understandings we carry with us into the analysis of the empirical, and thus we reduce the possibility of unreflective preconceptions which could unknowingly guide the analysis in specific directions.

Horizontal Analysis: The horizontal analysis is, as the name implies, designed to look across the family stories and draw parallels between the features which distinguish and separate the respondents. Basically, it means that the first part of the analysis extracts the concepts which the project wishes to develop. In the horizontal and cross-cutting part of the analysis, therefore, it is important not to confirm the preconceptions with new perspectives from the vertical analysis. The significance is on challenging the material with the knowledge which is continuously achieved by means of both theoretical and empirical surprises. This happens by, among other things, debating the points of reference from the individual interviews within the focus groups. In this particular case the points of reference of the horizontal analysis is an expression of the main mobility generating dynamics: *The pace of daily life, the importance of place and the creation of communities*. A clarification and gradation of these concepts adds to the development of nuances in late modern mobility theory. The three points of reference from the horizontal analysis thus ends in the final conclusion, as a subset of the overall recommendations which the project results in.

Conclusions based on a vertical and a horizontal analysis

By subdividing the analysis into both a vertical and a horizontal analysis, the individual voices attain their own place within the analysis. The vertical analysis shows the respondents' statements in their own context, structured according to points of reference. The vertical analysis is based on individual interviews. The horizontal analysis focuses on the characteristics and differences between the respondents. Here parallels are drawn to the theoretical concepts in order to identify the prevailing discourses. This part of the analysis is primarily based on focus group interviews, but also contains elements of the individual interviews in order to explain the context in which given problems are negotiated.

What this analytical frame provides is a thorough analysis of individual's wishes and dreams of everyday life, and the significance of mobility within this life. It creates a place for the voices of those whose mobility behaviour we aim to change, providing an opportunity to try to understand why it is so difficult to change. When working with vertical and horizontal levels in condensing empirical evidence, the process of revisiting the interviews and understanding the voices of each individual person often results in new and unexpected information coming forward.

The concurrent interplay between theory and the empirical, affects the conclusions drawn. The concepts and phenomena being tested are not static analytical orders to be tested in reality; they are open concepts which, through new inspiration from the theory and empirical data, are constantly developing. This development can only be described by clarifying the preconceptions of the research, the reflexive methodology, and the interpretative paradigm. Using this type of analytical frame increases indistinctness with the number of people interviewed, because of the massive amount of empirical work which then needs to be presented. In working with a larger group of people using both the vertical and the horizontal concept of analysis can be rewarding when structuring the analysis, though the final presentation will benefit from condescended versions.

Case 2: Everyday life of bus drivers – using interviews with action research

In this second case, individual interviews and focus groups are combined with action research oriented workshops. The empirical field is a link between the sociology of mobility research in cooperation with work environment research. The starting point for the research is the restructuring of the bus driver trade in Denmark. During the last 15–20 years, a restructuring, privatisation and outsourcing of bus lines in greater Copenhagen have occurred as a result of new public management strategies. This has led to increased pressure, stress, loss of breaks, and frequent shifts between bus operators. Time pressure, stress, a lack of flexibility and the loss of breaks are therefore crucial theoretical concepts related to the empirical practices of bus drivers' everyday lives. This project is based on researchers and bus drivers in Copenhagen working together in order to come up with suggestions as to how to improve bus drivers' working lives.

The research aimed to examine how specific characteristics of late modern life – also reflected in work life – are affected by mobility, and how a mobile work life affects mobility in other areas of everyday life. The project aimed to create opportunities of finding joint solutions to the problems experienced by the bus drivers, as well as avoiding an increasing individualisation of work and everyday life. The theoretical basis is the concept of 'flexible production' developed by Richard Sennett (1998), where he describes the negative consequences of the restructuring of working life during the flexible production regimes. Our study intends to go beyond Sennett's concept and intends both to include the subjective experiences of bus drivers in the regime of flexibility, and also to develop strategies and visions beyond flexibility – both at individual and community level (Drewes Nielsen and Aagaard Nielsen, 2006).

The methods used to create these new visions for working life, were action research, individual interviews and focus groups. Action research has a specific point of departure in relation to preconception, reflexivity and philosophy of science and can be defined as a participatory, democratic process concerned with developing practical knowledge in the pursuit of worthwhile human purposes. It is grounded in a participatory worldview which we believe is emerging at this historical moment. It seeks to bring together action and reflection, theory and practice, in participation with others, in the pursuit of practical solutions to issues of pressing concern to people, and more generally the flourishing of individual persons and their communities (Reason & Bradbury, 2008). Action research uses various methodological approaches, but our research is inspired by traditions of action research within the Scandinavian countries (Aagaard Nielsen & Svensson, 2006), with strong traditions of research collaboration between researchers and employed. In this specific case the research process is collaboration between the researchers and the bus drivers. The intention is that drivers themselves must add subjective experiences in order to qualify and clarify concepts such as time, place, communities and risk, articulated when they describe both their work life and everyday life. By using different types of qualitative methods, it contributes to a strengthening of the conceptual development and formulations of specific strategies for change in the future work of bus drivers. Again, several different methods are used to illuminate the same problem (Kvale, 1996). The following outlines two of the workshop methods used in the research: The future workshop and the research workshop.

Before the workshop, we conducted seven individual interviews and four focus groups, each with different categories of bus drivers (categories according to age, gender, ethnicity, experience, etc). The condensation, coding and analysing of the interviews, added the bus driver's subjectivity and experience to the concept of flexibility, and supported the transformation of the concept of 'flexible work' to the concept of 'volatile work'. Thus using different kinds of qualitative methods developed theoretical concepts essential in the research.

Workshop methodologies

The interviews created the platform for inviting and facilitating workshops with the drivers. Through these methodologies we – the research group – had gained in-depth insight into the bus drivers' everyday lives.

The Future Workshop: The different types of workshops stem from the future workshop methods (Jungk & Müllert, 1984). In this method, specific rules for communication and creativity ensures that the workshop is carried out on democratic and equal principles, with active participation of all involved. At the same time, the workshop establishes a specific learning space, where the participants' opinions and views are freely exchanged. The workshop is organised as a meeting away from daily routines in order to create a mental and physical learning space. It is inspired by different methodologies (Drewes Nielsen, 2006). Firstly, by action research approaches, where actors are involved and participating in a process of change; secondly, the workshop is facilitated and led by specific rules for communication with the aim of creating equal dialogues and eliminating the communicative power among the participants; and thirdly, the workshop facilitators obey specific rules for visualisations and creativity.

In the future workshop, participants – as the name emphasises – work with a forward-looking perspective in order to formulate specific drafts for the future. In this specific research project, the future workshop was entitled 'Our future life and work as bus drivers in the year 2017'. Ten bus drivers from greater Copenhagen participated. As part of the application of a range of methods, the same drivers were also interviewed individually as well as in focus groups.

The future workshop is divided into three main phases: a critique phase, a utopian phase and an implementation phase. In the critique phase, participants are asked to formulate negative criticism about their working lives. They were allowed to criticise all parts of a specific issue, although the other participants are not allowed to discuss the validity of others' critique. If a participant disagrees with a statement, he/she can pose a counter critique. Statements such as: 'breaks becomes shorter and shorter', 'it is stressful not to know one's working hours the following week', 'we are constantly changing managers' and 'the buses are worn out', are all examples of criticisms expressed and subsequently noted on flipcharts. The idea behind the critique phase is to give the participants the ability and opportunity to voice their frustrations. If the disadvantages of the bus drivers work life are outlined from the beginning, it is then far easier to create a free space among the participants later on in the process, in order to create fertile soil for constructive future scenarios.

The utopian phase is intended to be aspirational, where the focus is on wishes and dreams. Participants are now formulating utopian ideas of what would make everyday life as a bus driver more enjoyable, generally better and more sustainable. Dreams might include things such as: 'all drivers will have a thorough education', 'everybody helps each other', 'the work day is shorter', 'all the buses are running perfectly' or 'there is no uncertainty about future jobs'. Each of these utopian ideas is written on flipcharts as well. The foundation for the idea of the utopian phase is to stimulate the imagination of the participants. In order for the future scenarios to ultimately get air under their wings, and thus have a clear orientation against change, it is important to ensure there are no limitations caused by binding ties of reality and experience. All types of dreams are important and relevant within this process. The intended free space among the participants is a catalyst for new potentials for change. When the flipcharts are all full of utopian ideas, participants are then asked to vote for those utopias which they find particularly interesting. This is done in order to democratise and to focus the decision on what to carry into the wording of future scenarios.

In the third and final round, the implementation begins. Participants are asked to hold on to their wishes but must begin to make plans for how they can be fulfilled on a step-by-step basis. The participants join the working group for the utopia they find the most exciting, and these working groups are then asked to try to elaborate on the utopia and formulate proposals for action.

In between the three phases, small plays take place. Their primary purpose is to amuse, loosen the mood, start up the creativity, and get from one phase of the workshop to the next. In this way, the participants are disengaged from their growing thoughts, allowing them to open themselves up to new thoughts for the next phase. Another form of participant controlled creativity is introduced in the phases of visualisation, which is also scheduled in the workshop. These creative spaces are unpredictable oases in the densely packed programme. They challenge both the participants' and the spectator's frame of reference by means of humour, originality, involvement and thus a shared ownership and responsibility for the discussions as they evolve. After the future workshop, all statements are documented in a protocol, which is subsequently sent out to the participants. The aim of the protocol is to contain the memories of both the fundamental criticism and the utopian proposals, which we, as researchers, later on include in our further research analysis. The protocol is also the cornerstone of the organisation of the subsequent research workshop, which we, in this case, conducted half a year after the future workshop.

The Research Workshop: The reason why this form of workshop is given the term 'research workshop' is because it is an expression of the preconceptions, knowledge and experience, which the participants jointly hold in exploring new nuances in a specific field where each participant is an expert. The research workshop in this case was titled

'Future life and work as a bus driver in the year 2017'. The invited participants were bus drivers who previously attended the future workshop as well as new bus drivers, some of whom came from other companies. Also, a number of other 'experts' with various professional backgrounds – such as researchers, politicians and leaders of bus companies – also participated in order to contribute to further development of the future scenarios. The aim within the research workshop is to establish an equal and constructive dialogue between the two groupings. With the point of departure in the future workshop, the drivers presented their views on the problems as they formulated them in the future workshop. In this case, we, as the facilitators of the workshop, chose two future scenarios to be presented. Firstly 'a maximum six-hour working day', and, secondly, 'one united bus company on Zealand [the island around Copenhagen]'. Thus, a new free space is created through the establishment of a common understanding of the driver's terms of life and their values.

The research workshop creates a constructive dynamic process which is continually trying to reach further and deeper into the analysis of the drivers' practical reality. This is done through working together and sharing experiences with other professionals and research approaches. Simplified, we can say that this method aims to, in broad daylight, carry out a research process where the theory and the empirical voices mutually communicate and inform each other. The theoretical conceptual background informs, discusses and develops practical experience, and vice versa. The results of these workshop methods can be seen on several levels. First, in the form of concrete ideas for prospective changes which may be included in a planning process at appropriate levels and, secondly, in the form of contributions to the research process we, as researchers, work with.

Using different methodological approaches

In textbooks of qualitative methods it is often stated that triangulating of methods is one important way of securing quality in qualitative research (Silverman, 2005; Flick, 2006). In our research, every methodological step can be used separately, depending on the exact purpose of the research but, by integrating individual interviews, focus groups and workshop methods, different perspectives and knowledge is attained. In the individual interviews, focus is on how the individual explains, validates and makes sense of mobility. It reveals wishes and dreams, gives a thorough picture of what a mobile everyday life might look like, and what kind of individual ambivalences related to mobilities unavoidably merge with daily routines. In the focus groups, the explanations, the validation and the making of meaning are discussed and negotiated. Completing focus groups subsequent to individual interviews reveal the kind of explanations and validations which has a general accepted validity from other people and society as well. It also identifies how generally accepted truths are negotiated at a community level.

The workshop methods create a free space for dreams and wishes; the dreams and wishes which might otherwise not be accepted in the focus group. It creates a space which opens up for thoughts on 'how things could be different' regarding unarticulated dreams, wishes and coping strategies with great significance on everyday life mobility. Also, the interactive workshops bring new perspectives of future mobilities into the planning arena.

Concluding remarks

The aim of this article has been to show how mixing different qualitative methods can open up for the lived lives of mobilities. In understanding and changing mobility habits and conceptions we need to understand the role of mobility as experienced in the lived everyday life. By mixing methods we reveal different access points to the ambivalences and utopias and create opportunities for horizons of change. In the critical utopian interactive workshop, ambivalences and utopias are dealt with through common forums and solutions. By basing the workshops on qualitative interviews and focus groups we create a core of knowledge and the life utterances means that this is not just a 'game' detached from lived experiences.

This kind of research, working with a large amount of empirical material strongly calls for insights and reflections of theories and methods and of quality in qualitative research. Is has no meaning to transfer criteria of quality from quantitative research to qualitative research. Criteria such as reliability, validity and objectivity are not transferable to qualitative research. Qualitative research needs to find its own criteria for quality, including reflexivity, preconceptions, interpretative paradigm, transparency and research subjectivity. This is an ongoing discussion within qualitative research, and this article wishes to bring this debate into the emerging field of mobile methods.

We have illustrated how our own research has systematically been inspired by such reflections in two selected research projects of our own. These reflections have a two-sided aim. On the one hand, we, as researchers, are forced to make our analysis transparent, and thus we are forced to explicitly explain and justify our choices. On the other hand, with this transparency we produce research of a better quality, hopefully of great benefit to others for application and for inspiration. In that way, we have raised the external validity of our own research.

An important aspect of this reflexive work is simultaneously to seek clarification of our own roles as scientists in relation to the empirical field. One element is to clarify our own subjective driving force in committing the research, while another is to clarify the relationship we, as researchers during the process, have to the field of research. The qualitative methods, which are based on data production in the form of interviews, leave the researcher as the person making the subsequent interpretation of the empirical. The workshop methods and the action researcher, however, require a completely different direct involvement in the research field. Here, the researcher, in addition to the interpretative role, is also an active player in the processes of change in the research field. Thus the workshop method makes the normativity of the researcher more obvious and challenging than when working with interviews.

Mobility research is a developing research field, both theoretically and empirically. New concepts are to be developed, and new empirical studies will hopefully join, intervene and contribute to the theoretical development. The tacit knowledge of experiencing mobile life has an important future role to play in qualitative research to be carried out in this emerging field of research. These future research practices can hopefully contribute to our common understanding of conflicts and ambivalent views on the transformation of modern mobile life. The ongoing discussions and present practices of quality in qualitative mobility research can play a crucial role in this development.

3
In-vivo Sampling of Naive Drivers: Benefits, Practicalities and Ethical Considerations

Ian Walker

Anybody who applies principles from quantum physics to everyday life is either in the business of selling quack medical cures, or plain wrong. As I am selling nothing, I know I am making a mistake when I say studies of driver behaviour often make me think of Werner Heisenberg. As you may know, Heisenberg argued that the very act of observation can change the thing being observed. Although he was concerned with the weird world of subatomic particles, and what goes on there does not apply to the large-scale world where cars, pedestrians and other such interesting things live, it is still intriguing for us as researchers to ask whether a similar principle might apply in traffic analysis. Could our attempts to study people's traffic behaviour be undermined because our observation changes the very behaviour we wish to study?

In discussing their recent meta-analysis of driver distraction from mobile telephones, Caird *et al.* (2008) make the point that the relatively substantial effects seen on distracted drivers' reaction times almost certainly represent an optimistic position, as these were drivers who were fully aware they were involved in research, and so were almost certainly acting to the limits of their abilities. In the real world, Caird *et al.* suggest, the impact of dual-task performance on driving is almost certainly greater than seen in the laboratory, or even in observed on-road driving.

And this brings me to the point of this chapter. Road accidents tend not to happen in scientific studies, they tend to happen during people's everyday lives. This is partly because accidents are fairly rare, and the amount of driving which is observed by researchers is minute compared to the totality of car-use, but also because it is entirely reasonable to believe that people will not behave the same way during a scientific study as they do in everyday life when it comes to some of the behaviours likely to contribute to accidents such as driving too fast, crossing a road in a dangerous place, and so on.

The first of these concerns – that research studies simply do not measure enough driving to capture the relatively rare events which lead to accidents – was the rationale behind the Hundred Car Study in the US (Dingus *et al.*, 2004).[1] Here 100 cars, driven daily by various people, were fitted with sensing and recording equipment such that over time it was likely multiple accidents and other events of interest such as near-misses would be captured for study. The project was successful, and highlighted in particular the role of driver distraction in road accident aetiology.

However, it is still a matter of debate the extent to which the drivers in this study might have had their behaviour altered by the knowledge they were under constant observation. Certainly, the fact so many people were recorded having accidents or near-misses whilst eating or attending to personal hygiene might suggest they soon forgot they were being watched, but it is impossible to know whether the number of such incidents would have been still higher without the data recording. At present, to the best of my knowledge, we do not have a definitive answer on whether a person's knowing they are under observation definitely changes their driving behaviour, but the idea has such *a priori* plausibility that this chapter will proceed on the assumption that it does, and that we can gain benefits when we seek to understand normal driving behaviour if we are able to observe naive drivers, i.e., those who do not know they are under observation.

I used this general approach of testing naive drivers in my 2007 study of influences on drivers passing bicyclists (Walker, 2007). Here I took a bicycle equipped with data recording devices and spent several weeks cycling around English streets, recording how much space passing drivers left between their vehicles and the bicycle, as well as various properties pertaining to the overtaking event such as the bicycle's position on the road, and my appearance at the time (e.g., whether or not I was wearing a helmet). With a database of over 2300 overtaking events I was able to find some intriguing influences of my appearance and behaviour on the safety margin left by drivers. The fact there were reliable relationships between my appearance and behaviour and how much space passing drivers left led to a useful conclusion: when drivers overtake bicycles they do not simply do this the same way each time; rather, the passing behaviour appears to be tailored to some extent to the individual circumstances of that moment. Moreover, the nature of this behaviour adaptation seemed quite clearly to reflect certain common misunderstandings about bicyclists such as the belief (e.g., Basford *et al.*, 2002) that helmets are a sign of experience and skill.

There are other methods available for studying drivers' everyday behaviours, many of which are more passive. Driver observation can be as simple as posting observers at roadsides to record the behaviours they observe, to video recording either at static sites or from moving vehicles immersed in the traffic. Let us first consider some of the practical issues surrounding such research before moving on to the matter of ethics.

Practicalities

Perhaps the first issue to be considered by anybody planning to record people's road-use behaviour is how the observations will be taken. And perhaps the first issue here is whether observation will be overt or covert.

This issue is perhaps best illustrated by the work of a US transport researcher tasked with conducting regular surveys of seatbelt use in his state, where their use was compulsory. He employed researchers to stand by traffic lights; when the lights turned red and a queue formed, the researchers would walk along the queue noting how many people were wearing seatbelts and how many were not. However, the project ran into difficulties owing to the fact seatbelt compulsion was quite a politically charged issue. Often, as soon as people became aware of the observers, they would deliberately unfasten their seatbelts, presumably to make a statement about their views on compulsion; others would quickly fasten their belts. As such, the researchers soon learned they could only obtain data from the first two or three cars in any given queue, making their task much more labour-intensive and time-consuming.

A solution might have been to make the observation covert. This would have raised some ethical concerns over privacy, as we will see below, but it also introduces practical problems. Few employers in these safety-conscious times would consent to their employees standing by a roadside without high-visibility clothing, which is clearly not conducive to covert observation. Perhaps the best solution for covert observation, then, would have been a remote-controlled video observation system, but this makes studies considerably more expensive. At the very least, it adds the cost of the electronics to the cost of the human observers (assuming they record the data in real-time); more likely it increases the cost of the study more than this: if the data are recorded then analysed at a later time there is the cost of the electronics, the people to operate the equipment *and* a whole extra labour-intensive process of transcription later.

Another important practical issue with recording data from drivers concerns our ability to generalise whatever the observations tells us, in part because it is inevitable there will be a lot of noise in the data. Because human behaviour is so variable, it is preferable when testing people to test many individuals and to test each individual many times, that is, we obtain multiple measurements from each person and collapse these to a single score, reducing the influence of any random factors which might influence behaviour moment-to-moment. We then further collapse the individuals' already collapsed scores to provide an average measure of performance, as a proportion, a mean/median or as a coefficient in a regression model.

However, in almost any sort of observational study where drivers do not know they are being observed it is not possible to test each person many times, and so each person can only provide a single datum. This will lead to noise in the data that would have been eliminated in a more typical procedure.

A related concern is that each observation has considerable temporal and spatial limitation. Specifically, in almost all circumstances, whether using observation at a fixed point or a mobile system incorporated into a vehicle, we must be aware we are sampling only a very small fraction of each person's journey, and that each observation is intimately tied to the place at which it happened. Roadside observation might tell us very little about people's behaviour in circumstances different to those in that particular location; mobile on-road observation might become more varied than we would like because each behavioural observation is 'coloured' by the exact location at which is was made, all of which are different. These influences mean any data we collect will almost certainly be noisy and any models we construct from them must always be viewed with some caution.

About the only practical step we can take to address the concern about temporal limitation – the fact we are sampling only a fraction of each person's journey – is to ensure we collect as many observations as possible to try to average out individual differences and to try, across many people, to cover various aspects of people's journeys.

With regard to the spatial limitation we have two options. We can either collect our data at a very limited range of locations (or even just one) or go on-road with mobile recording and collect it at many locations. Both approaches have strengths and weaknesses. If we collect data from a limited range of locations we can eliminate some of the noise in the data but must be cautious about generalising our findings to other circumstances than those we studied. If we collect data using a mobile arrangement any findings will be less tied to a given location and so are perhaps more generalisable, but at the same time, as mentioned above, we risk the data being noisier owing to many diverse situations being pooled together.

As a final point with generalisation, we need to be conscious of the fact anything learnt from observation of drivers on one country's roads, even when the observations are from many people at many diverse locations, may still not generalise at all to other countries, where the culture and traffic system is different.

I have raised a lot of concerns here about potential problems with collecting data from many different individuals, but there is one potential advantage. Many statistical procedures make the assumption that all the data points to be analysed are independent of one another. If each data point comes from a completely different person – and perhaps even from a completely different location – this assumption of independence should be easy to fulfill, which is one less concern. (Although might drivers following one another be influenced by seeing the person in front?!)

As a final very practical consideration, you should note that in many locations it may be necessary to get permission from a local authority, highway authority or the police before conducting any sort of observation. This sort of requirement clearly varies so much from place to place all I can do is suggest you check in good time.

Ethical issues

Perhaps the most difficult aspect of naive driver observation is the ethical dimension. There are several issues which need to be considered, and I am not in a position to provide hard-and-fast rules here as ideas of what is and is not acceptable vary between individuals and between research institutions. However, I hope that by highlighting some of the main issues which should be considered I can provide some useful planning guidelines for any researcher considering observation of naive road users, particularly for people from backgrounds such as, say, engineering, where issues of human testing might not be as pervasive as they are for us psychologists.

The first issue is one of consent. It is a fundamental plank of human research that wherever possible participants must be able to give informed consent to take part in any study, such that they take part knowing what the study involves and what possible risks they face through their participation. Clearly, informed consent of this sort is simply not possible for the type of research we are considering here.

The extent to which this matters will depend upon the type of study being conducted. A project which simply involves observation of people's spontaneous behaviours is relatively uncontroversial in this regard. Provided the behaviour takes place in an obviously public space, it is difficult to see a strong objection to anybody watching it. However there are clearly gradations to be considered. Is behaviour carried out in a quiet residential cul-de-sac really as public as behaviour in a busy city centre? And certainly behaviour on the margins of public and private space – such as the observation of people's parking behaviour in their driveways or roadside garages – is more questionable and should be considered carefully.

Intimately connected with the issue of observation is the issue of recording those observations in some way. If I drive down a public street then I cannot reasonably object to your seeing this, but I might be able legitimately to object to your recording my journey – especially if this could be construed as a form of covert surveillance – if I knew about it.

As such, perhaps a useful guide to our work as researchers can come from the degree of anonymity inherent in the data we collect. Overpage is an example of a sliding scale of data and privacy. Let us imagine we see a red car driving down a certain road on a certain day at a certain time, and think about various ways we might observe and record this event.

Clearly this does not cover every eventuality, but should make clear the fact there are many possible levels of invasiveness in the act of observation and recording. Some of the possibilities listed above would allow the clear identification of the driver and would provide a record of their exact movements, whereas others would not.

As well as the level of invasiveness inherent in the data collected one must also consider the level at which the data will eventually be reported. Some would argue that if the behaviour is in a public place it is 'fair game' for observation and recording of all forms. Others would argue that the level at which the data are reported is the prime ethical concern for a researcher: for example, many researchers would feel it is ethically acceptable to make a video record of a public place provided the data are only reported at one of the less invasive levels which does

No observation at all   Least invasive

Observation, no data recorded

Including the car in an overall vehicle count

(e.g., '50 vehicles passed')

Including the car in a divided vehicle count

(e.g., '20 red cars')

A record involving time and location (e.g., 'a

red car at 1210 on 7 September on Standish

Street')

A record involving specific characteristics

(e.g., 'driven by a white middle-aged man')

A record involving unique characteristics

(e.g., vehicle license number)

A photographic record

A covert photographic record

A time-stamped covert video record Most invasive

not allow the identification of individuals (e.g., the video is used to produce an overall vehicle count). Incidentally, 'reporting' can mean many things, from technical reports to journal articles to television shows. Generally, I would suggest, it would not be appropriate, with data collected through naive observation, to present any data which could identify an individual in any forum at all, even amongst fellow professionals. So no hilarious video footage of people's bad driving at conferences or seminars!

Finally, before we move on from data collection, I should also mention the issue of data storage. If you collect any sort of data at all – with or without the consent of the participants – you should seriously consider keeping anything which might allow an individual to be identified under lock-and-key, or under password protection if electronic.

So far I have considered issues surrounding passive observation: issues surrounding data recording and reporting in studies where we simply observe naturally occurring behaviour. However, there is a whole other class of observation studies whereby the researcher somehow influences the drivers' behaviour in a way that would not have happened had the research not been taking place. I faced this issue when I carried out my bicycle overtaking study: had I not been conducting this study I would not have been cycling on those roads at those times and over 2300 overtaking events would not have occurred. The research process was changing people's behaviour from what it would otherwise have been. Why does this matter? One reason is that each of these events carried a risk of collision. Had a driver killed or injured me, this would have had serious consequences for that driver's wellbeing. They could even have ended up in gaol! This risk to the drivers' wellbeing would simply not have existed had I not conducted the study.

In this case the risk was, I felt, justified on several grounds, including the value of the data to be obtained and the fact there was nothing unusual about a driver overtaking a bicyclist on a city street. However, what if the study had been recording drivers' behaviours around a novel experimental traffic engineering intervention, which they could not reasonably be expected to handle as a matter of routine?

In a related vein, in any sort of study where the researcher carries out an intervention or experimental manipulation that could conceivably harm the wellbeing of participants, it is important that the study methodology be sound. If not, the researcher is risking people's wellbeing for worthless data. For example, if a researcher were to change a road layout and record driver behaviour after the change, without also collecting observations before the change took place, or some of other suitable control data, the study would be methodologically flawed: with only behaviour data from after the intervention there would be no way of knowing whether the intervention had altered the behaviour or not. As such, the risk to the participants had no value, and the study would usually be considered unethical.

The final major ethical issue we must consider, which perhaps falls into a very different category to all those above, would be what to do if a researcher witnesses and records an instance of illegal or dangerous behaviour. Let us say you have fitted a vehicle with a video camera and recording devices to record some sort of behavioural data on the road. What if your record included an act of illegal driving *and* sufficient information to uniquely identify the perpetrator? Should you, in this case, provide the evidence to the police? The difficulty comes from the fact the person did not consent to have their identity or behaviour recorded in the first place. This would be particularly problematic in circumstances where the driver's behaviour might possibly have been a response to the existence of the study – as in the example of the experimental traffic engineering project discussed above.

As I hope you can see, the process of studying traffic through observation is filled with contentious ethical dilemmas for the researcher. There are no clear answers or definite guidelines, but the wise researcher would consider all these issues, and have a stance on them, before embarking on any study.

Final thoughts

In vivo recording of data from road-users who do not know they are being observed has an important role to play in our understanding of road use and safety. The possibility will always exist that people who know they are being observed will alter their behaviour as a result, and naive observation seems the only realistic way to address this. Here I have highlighted some of the key practical and ethical issues that a researcher undertaking such a study should consider.

As a final point, I should emphasise that I am in no way suggesting that knowledge of being observed invalidates the bulk of studies using on-road testing, driving simulators and so on. These are perfectly valid methods for investigating a whole range of topics. If, for example, a researcher wants to understand issues surrounding perception, attention, decision-making and so on, naive observation will probably have little to offer whereas these more controlled methodologies do. Moreover, naive observation is useless for studying a whole range of other interesting areas: attitudes to traffic and transport, the level of drink- and drug-driving and so on.

Where naive driver observation really comes into its own is for studying overt behaviours where there might be any sort of social or moral imperative for people to suppress or alter their actions when they know they are under observation: speeding, careless driving, inconsiderate behaviour towards other road-users, and so on. For topics such as this, where social pressure means controlled studies are unlikely to reveal how people naturally behave, the technique has a great deal to offer. Ideally, it would be supplemented with further work to explore precisely *why* people would want to show a disjunction between their public and private behaviour.

Note

1   For a useful summary of this study see http://www.docstoc.com/docs/805732/An-Overview-of-the-100-Car-Naturalistic-Study-and-Findings

4
Narrating Mobile Methodologies: Active and Passive Empiricisms

David Bissell

Introduction

Work on mobilities – as a collective interest of many working in human geography, sociology and cultural studies – is one such area where methodological transformation is currently high on the agenda, reflected in the number of workshops and conference sessions dedicated to this topic (RGS-IBG, 2006; RGS-IBG, 2007; CeMoRe, 2006). In light of recent calls for social scientists to work more closely with their objects of study (Latour, 2000 in Gane, 2006), many researchers argue that a new range of research practices need to be developed which attend more successfully to the experience of being on the move. Whilst I am broadly in agreement with the argument that we need to develop more creative 'mobile methods', in this chapter I want to take a step back by considering, and critically reflecting on the type of research that emerges from the development of these methods. More specifically I want to suggest that these methods, which often privilege particularly active dimensions of the mobile body, may be less well-equipped to get at and narrate some of the less-agentive experiences of mobility, where bodies are pacified and not engaged in any form of intentional, auto-affective action.

This chapter begins by describing some of the challenges that mobile methods generate, suggesting how certain mobile methods may run the risk of privileging a particular style of narration that potentially eclipses a range of phenomena which might be equally integral to the experience of movement. I proceed by outlining some of these experiences which are characterised by their passivity, or *quiescence*, such as lethargy, tiredness, hunger and pain. I suggest that these might be hugely significant experiential dimensions of mobility but might be overlooked by our current methodological toolkit and textual strategies with which such experiences are narrated. The final section hints at some of the ways by which mobile methodologies might be developed to become more attuned and responsive to these phenomena.

Mobile methods, active empiricisms

The proliferation of mobile methods has been instrumental in highlighting some of the various permutations of what it is to experience movement. However, I want to suggest that some of these methods run the risk of privileging a specific type of narration that obscures and potentially negates a range of phenomena which I will argue are integral to the understanding of what it is to experience movement. Whilst acknowledging their huge diversity, what perhaps links them is their tendency to focus on particularly *active experiences*. To mobile researchers, what are important are not only the active speech patterns and streams of active consciousness that might be iterated through dialogue, but the various *gestures*, *actions* and *movements* of people. The key point here is the way in which people and objects on the move are narrated and accounted for in a way which privileges action and activity over other perhaps more fragile ways of being mobile. As Kusenbach herself admits, 'ethnographers take a more active stance towards capturing their informants' actions and interpretations' (2003: 463). It is these actions and interpretations that emerge most compellingly through narration. Put simply, we need to be careful that such methods and more specifically, the narration of such methods, are not celebrated as the new orthodoxy with which to interrogate and understand mobilities, but rather are critically incorporated into a more mature range of methodological techniques. At the risk of masking differences between approaches, there are three broad interrelated reasons why particularly mobile methods might pose a critical problem by obscuring some of the possible ways through which movement could otherwise be narrated and consequently theorised.

First, amongst some accounts, there is a dilemma of *epistemology*: specifically the way in which mobility is apprehended and conceptualised theoretically. Though the emphasis of much social science research practice is increasingly inductive, the manner in which some researchers *approach* the challenge of mobile subjects might actually be a rather more deductive practice. This has significant implications for the ways in which these mobile subjects are construed. The heralding of a new mobilities paradigm has been instrumental in recognising the significance of thinking through ways in which interrelated movements permeate everyday life. Yet the very establishment and recognition of such a paradigm may have the effect of prioritising movement when approaching objects and subjects, bound up in the *a-priori* way in which subjectivities are conceptualised. Rather than thinking about figures that have the potential for movement, the 'mobile subject' is *defined by movement* and always-already on the move. Subjectivities are mobilised prior to empirical investigation through this act of paradigmatic naming. Put simply if we set out to investigate *mobile* environments, the ethnographic gaze might be orientated and focused far more sharply on these shifting, fluctuating *moving* dimensions of mobility (see Laurier, this volume, for further critique). This is perhaps not surprising given the current trend of theorising objects and subjects through metaphors of movement. In addition to Cresswell's (2001) nomadic metaphysics, contemporary mobilities research has been heavily influenced by the theoretical poststructuralist semantics of 'liquid modernity' (Bauman, 2000), 'fluid objects' (Law & Mol, 2001; Urry, 2003), 'transitory geographies' (Crang, 2002) and 'spaces of flows' (Castells, 1996) to name but a few. Following this logic, whole landscapes have been apprehended as mobility spaces, transient spaces or non places (Augé, 1995): its most familiar rendering condensed in the iconic international airport (Crang, 2002; Hannam *et al.*, 2006). The myth of unrestrained, deterritorialised movement has been written about by many (see Kaplan, 1996; Cresswell, 2001 for example), particularly the various ways in which mobility is socially differentiated (Massey, 1994). However, the way in which such theorisations are achieved often set movement in opposition to non-movement, stasis, or 'moorings' to use Urry's (2003) term. Such differentiation of mobility clearly heralds a certain maturity in thinking through the implication of the complex relationalities between mobile subjects and objects (see also Adey, 2006). Yet to set up a dialectic, however relational, where what is other to mobile is 'immobile' or 'moored' might be problematic in that all subjectivities are defined and evaluated according to the logic of *movement* which reproduces and amplifies its rhetoric. Such a dialectic might not leave room for thinking through relations to mobility that transcend such a mobile apprehension.

Secondly, there are issues of *ontology* that need to be considered. This relates to the ways in which bodies caught up within these mobilities have been conceptualised. This in part follows a move within the social sciences which have seen a renewed commitment to practice-based theory and the promise of performance (see Thrift, 2008; Nash, 2000; Harrison, 2000 amongst others). Such performative ontologies have influenced many researching mobilities and have consequently prompted a change in the way through which mobile phenomena is conceptualised (see Bærenholdt *et al.*, 2004 in relation to tourism mobilities), emphasising the 'felt, touched and embodied constitution of knowledge' (Crang, 2003: 501). This emphasis on the 'more-than-representational' (Lorimer, 2005) dimensions of experience that fall outside of systems of meaning and signification increase the attention to the corporeal sensuality of certain mobile practices: how bodies are folded through and shaped by landscapes during the course of mobile events (see Wylie, 2002, 2005). However, this commitment to liveliness and the body-in-action also has further implications for our understanding of the experience of mobilities. In making subjects 'dance a little' (Latham, 2003: 2000), such performative renderings of mobile subjects, in part inspired by various strands of vitalism, might have the effect of generating an *overanimated* mobile subject. This potential for movement and connection that privileges the body-in-action, as active and agentive (see McCormack, 2002, 2003), might therefore be construed as 'overanimated' in the sense that active movement neglects other corporeal subjectivities. These performative methods have arguably expanded the range of methodological resources available to those researching mobile bodies however these more diagrammatic styles developed tend to emphasise the creative corporeal kinaesthetics where focus is on the 'energetics of life' (Kuppers, 2000: 135), of movement as active, ongoing achievement. Ethnomethodological approaches to the study of mobilities similarly focuses on only a narrow range of what it is to be mobile, which privileges speech patterns, bodily movements, gestures and placings (see Laurier, 2004). Whilst such apprehensions of an active, agentive rendering of subjectivity can act as a proxy for various affective complexes, such mobile methods might only capture a narrow spectrum of what it might be to experience these mobilities. To attend to and emphasise the sensuous and more-than-representational *performative* dimension of mobile experiences could potentially serve to reduce the rich variety of subjectivities to the various permutations of kinaesthesia, the 'reeks and jiggles' of movement (see Hutchinson, 2000).

Third, and linked to this, are issues of *practice*. This is about the range of empirical and narrational strategies that might be used to attend to the mobile experience. With the development of more mobile methodologies, the vocabularies enlisted to narrate these experiences are similarly active. This might be symptomatic of the restricted range of dissemination channels available to academics to communicate their research which tend to privilege textual material. Academic prose is rather more suited to capturing and narrating a certain necessarily limited range of experiences and often has the effect of creating 'very wordy worlds' (Crang, 2003: 501). As Crang (2005: 230) comments 'qualitative research, despite talking about the body and emotions, frames its enterprise in a particular way that tends to disallow other forms of knowledge'. Notwithstanding these restrictions, this problem is exacerbated by the type of narration these words have the capacity to create. Through the messy filtration processes whereby practical doings are translated into linear prose by the researcher, the massive, complex, undulating and often contradictory experiences of movement are stripped and squeezed into piecemeal fragments for digestion. Whilst the limitations of verbal and text-based sources have long been recognised, it seems that few methods are immune from this problem. Narrating mobile methods such as 'go-alongs' that feature *moving* as an active dimension are similarly poor at translating and accounting for dimensions of experience that are not either active or intentional. The narration of these mobile methods tends to privilege the *active* dimension of corporeal experience: I *walked*, I *ran*, I *watched*, I *talked*, I *remembered*. Put simply, it is perhaps easier to talk about activity than inactivity and therefore such narrations clearly capture only a restricted presentation of the mobile body. Frustratingly, there seems to be an aporia when the body is not engaged in activity as testified by Watts in her mobile ethnography of train travel: 'Nothing is happening in this carriage...Time and place here are static and unchanging. No one is doing anything – barely moving – as if in cryogenic freeze or stasis...Nothing seems to happen...I want to write that something happens. But nothing happens. A man reads a book, then reads a newspaper. A woman fidgets and sniffs...' (2008: 722). Experiences and sensations that cannot be reigned into the systemisation, thematisation and conceptualisation of social research seem to be neglected (Harrison, 2008).

Each of these three issues suggests that we need to consider a little more carefully about what we are actually producing when we use mobile methodologies. Whilst the diversity of methods and theoretical underpinnings means that it would be disingenuous to generalise, it might be useful to consider how different mobile methods reveal a particular series of tensions which might relate to one or more of these issues. For example, go-alongs and co-present mobile ethnographies (Kusenbach, 2003; J. Anderson, 2004) might reveal significant interactional textures, but they might obscure those dimensions of experience that transcend individual bodies, through a particular commitment to logocentrism. Conversely, experimental performative mobile methods (Wylie, 2005; McCormack, 2003) which problematise the figure of the body might be attuned to the post-phenomenological melding of body and landscape, but they might equally serve to 'overanimate' the body. More significantly, however, is that the grammars and vocabularies used to narrate these mobilities often tend to privilege activity and intention over other ways of experiencing mobilities.

Passive mobilities

Where both the practical undertaking and subsequent narration of mobile methods might hone our attention to particularly active dimensions of mobilities, in this section I want to suggest that there might be a range of other ways in which people experience or relate to mobilities which might not be so easily attained through these methods. More specifically, these mobile methods might not be adept at capturing those aspects of experience that are less-active and less-agentive that are integral to the everyday quotidian experience of being mobile but might be overlooked by these methodological techniques. Such experiences might include weariness, tiredness, lethargy, hunger and pain: all hugely significant entanglements of body and world that might be precipitated through movement but that seem to dodge the lasso of many mobile methods. Indeed much of the duration of moving through mobile spaces such as airports might be spent 'doing' very little, where the body is involved in the quotidian, less-than-agentive rites of passage implicated through waiting and queuing. In different, less privileged circumstances, we might think of other relations with movement which might be characterised by a similar sense of the unagentive such as people subjected to extraordinary rendition (Spark, 2006), or how whole populations are forced to live in limbo (Bayart, 2008). In these situations characterised by passivity, the body is more inert, inactive and is not so involved in the ongoingness of performance. Whilst some forms of limbo might be highly active and agentive (Bissell, 2007), what I want to focus on here are the suspensions that 'trace a passage of withdrawal from engagement' (Harrison, 2008: 424), where the body bows out of the schema of auto-affective action. A few writers have documented the fatiguing effects of travel such as Smith (1995: 1441) who recounts that 'the fatiguing effect of travel seems to apply even under apparently ideal conditions: travelling for pleasure, first class, by rail, no driving either end, good companionship, no anxiety about possible delays. Still it is tiring'. Similarly many travel writers have documented the wearying effects of movement on the transported body (see Diski, 2006 for example). Indeed the relative absence of these passive or *quiescent* experiences of mobility within social scientific literature might therefore be surprising, even startling.

Such a commitment to corporeal quiescence reflects a nascent body of work within the social sciences that has interrogated the sensibilities of the stilled body (Bissell & Fuller, 2009; Harrison, 2009; Murphie, 2009). Harrison's (2008: 425) intervention focusing on passionate desubjectification outlines a way of thinking through this stilled body as an event of the unwilled where 'synthetic activity does not happen, when the everyday flow and exchange of meaning stutters and abates and goes awry'. Similarly B. Anderson's (2004: 744) work on boredom centres around some of the various affective states of stilling and slowing time, where the affective energies and vitalities of life are suppressed. In a related vein, Thrift (2006) has recently alluded to the importance of thinking through various forms of mesmerism relating to a form of semi-conscious being or a suspension of disbelief. Others have been interested in charting the sociality of quiescent experiences such as sleep (Williams, 2002, 2005; Williams and Boden, 2004). However these interventions are characterised more by their (often dense and perplexing) theoretical negotiations, with few researchers suggesting how or even whether such experiences can be empirically investigated. Indeed through work on sleep (for example, Kraftl and Horton, 2008), there is often a tendency to apprehend experiences of quiescence as embodied social *activities*, thereby translating these experiences once again into active sensibilities.

There is a danger of semantics here though in aligning passivity, inactivity or even immobility with physical stasis. Many researchers have explored the various relationalities between the movement and stasis of bodies and objects (Hannam *et al.*, 2006, amongst others). Contrary to those who have romanticised about unimpeded nomadic movement, these tracings have been far more attentive to the ways in which mobilities are differentiated and constituted as much by immobilities or moorings as by movement (Cresswell, 2001; Urry, 2003). As such, thinking through 'immobilities' might offer a more responsive way through which movement can be theorised. Not only do certain bodies move faster at the expense of other (more slowed) bodies (Gottdiener, 2000), but various assemblages of stilled objects in position are required for other objects to move, such as the massive infrastructure assemblages that make up airports (Adey, 2006). Similarly, Adey (2007) describes some of the various practices of spectatorship which serve to temporarily 'hold' attentive bodies in airports. However even whilst being held, enduring physical stasis, these bodies are engaged in auto-affective activity: watching people, looking through windows, window shopping or attending to information screens. Similarly the verbiage here relies on active and rather traditionalist renderings of power and agency of *being held* where the holding of the body implies the coercive forces of pressure and tension.

Whilst hugely informative, narrations of these rather more active practices tell us very little of the construction or destruction of various types of subjectivity; active or quiescent. For example, relative movements and stases as *simultaneous* experiences are overlooked through much of the recent body of work on automobilities where the onward movement of the car is juxtaposed by the incarcereal holding of the body in the seat (see Featherstone *et al.*, 2004). Pushed further, one of the most pressing experiences of driving a car might be the ensuing quiescent experiences of weariness and exhaustion. Yet these are experiences that are often disregarded. This might not be surprising considering that a large number of these studies are informed by various contemporary renderings of actor-network theory (see Urry, 2007). Whilst this draws attention to some of the complex, distributed materialities that make up these intersecting mobile systems, a problem with such an approach is its refusal to consider the complex nature of generative subjectivity and the implications of various forms of movement on embodied experience. To apprehend movement from a more post-phenomenological perspective is to view these events of quiescence as a specific kind of relation-to-the-world that transcends and folds through this relational dialecticism of mobility and immobility. To consider bodies through a polarised lens of movement and stasis refuses to acknowledge the complex ways in which bodies can bow out of this schema through the fleshy corporeal entanglements with movement.

Writing about passivity within social science is of course not new. Many writers interested in thinking through the various shapings and implications of power geometries draw on the verbiage of passivity to describe a particular and often subservient relationship to a dominant power (Massey, 1994). For example, postcolonial research has long been interested in the complex power disparities between colonialists and those colonised where the latter is commonly described through discourses of passivity (Spurr, 1993). Such power inequalities have also been considered as emergent from the relations between a variety of human and non-human actants. This has been particularly explicit in research that has investigated relations between new technologies and bodies – such as how mobile surveillance and camera technologies can work to pacify and subdue bodies (Graham & Marvin, 2001; Crary, 1999). However, there seems to be little room in these studies for weaving through a more positive (but crucially not productivist) form of corporeal quiescence. This may be partly due to the way in which social scientists, and human geographers particularly, have often privileged a particular rendering of agency where a lack or absence of agency can only be considered to be undesirable. Recent insights from performance-based approaches have demonstrated well how power and agency are in fact more fluid, fractured, multifaceted, and crucially enacted conditions (Thrift, 2004; Amin, 2004; Allen, 2003). Such approaches have been instrumental in illuminating how power and agency are not 'held' objects, but emerge through performance. However, to consider passivity and corporeal quiescence in this schema still renders it as somehow undesirable: to lack agency is still to lack power. This move towards thinking through more seriously the nature of stilled bodies highlights how subjectification remains one of the difficult yet crucial aspects of social enquiry to grapple with. Furthermore one could ask whether it is indeed responsible to talk about passivity at all. Are they aspects of experience which because they are non-agentive, we should not concern ourselves with? Whilst some may see a shift towards thinking through the various forms of passivity and susceptibility of bodies as an apologetic expression of repentance against the often masculine power*ful* geographies of the past, thinking through the implications of desubjectification is crucial in developing a more balanced view of subjectivity: subjectivity as responsive rather than always-active.

Such a nuanced conception of subjectivity has far reaching implications on the ways in which researchers characterise mobile bodies. To take seriously the passivity of bodies is crucial when considering the embodiment of various time-spaces (May & Thrift, 2001). The relationship between passive bodies and temporality may be very different to active bodies. When bodies bow out of auto-affective action perhaps through tiredness or hunger, their temporal sensibility shifts and alters. Similarly such corporeal stilling can have profound effects on the ways in which spatiality is conceptualised, perhaps as becoming more enclosed, claustrophobic or alternatively more open and expansive. Additionally where bodies acquiesce may have a distinct spatiality, for example drawing on a train journey, certain places may serve to heighten a non-agentive sense of quiescence such as a waiting room (Bissell, 2007). The types of subjectivity in part fostered by a particular environment or a particular space is therefore of interest also to designers, advertisers and engineers when planning such spaces where types of susceptibility can potentially be woven into the fabric of the space.

This model of subjectivity that takes seriously forms of withdrawal from engagement enables us as researchers to interrogate those aspects of experience that emerge from and are integral to the rite of passage but have been traditionally overlooked. Bachelard's daydreaming demonstrates well how this form of quiescent subjectivity enables us to think through the unwilled slides between the real and imaginary – experiences that 'transcend the surface categories of immediate common-sense experience' (Picart, 1997: 69). Whilst on the move, the dominant mode of being in the world may not necessarily be one of sustained, engaged activity. In a society which is characterised by the multiple demands of increasingly longer commuting times (Koslowsky *et al.*, 1995), extended working hours together with balancing the tight-rope of family, work and social life, the study of these non-agentive quiescent sensibilities such as tiredness and lassitude seems more pertinent than ever before.

However, the relative absence of such experiences in studies of mobility points to a problem of both theoretical and methodological significance. Whilst theoretically, passivity is creeping up the agenda, crucially it would seem to be during the initial empiricisms that these passive experiences whilst enacted *are not registered* or considered significant since our current empirical toolkit is skewed towards cap-turing a reactive sense of subjectivity. Whilst acknowledging their the-oretical significance how then can we begin to research these moments of corporeal susceptibility? Is it possible to construct a more responsive empiricism that is more responsible to these phenomena that transcend both movement *and* stasis?

Narrating passivity

There have been a handful of studies that have looked at corporeal quiescence empirically. Such studies have emerged from the field of medical anthropology and have tended to focus on the various experi-ences of corporeal stilling through the event of waiting. Specifically these studies have investigated some of the various affectual complexes experienced by people in waiting rooms (see Bournes & Mitchell, 2002; Brekke, 2004; Buetow, 2004). Whilst responding to the absence of empir-ical material on what is arguably one of the most commonplace events not only in primary care but also through the event of travel, these studies have tended to focus on the active dimensions of the waiting experience, of what the body *actively does* during these periods argu- ably at the expense of other ways in which bodies experience these durations. Again, active verbiages are enlisted as a strategy to construct a narration of waiting that is full of energy, vibrant, shaking and above all, *intense*. Some of the complex, and perhaps to be expected, emergent affects experienced by bodies *are* prised out, such as fear and anxiety (Bournes & Mitchell, 2002). Yet other sensibilities such as lethargy, indolence or sluggishness are neglected. Other studies which have attempted to attune to the experiences of stilled bodies are highly quantitative (see Buetow, 2004) where embodied experiences are causally (and rather problematically) inferred from duration.

So how best are methods that remain responsive to this range of experiential affectivities, not only of the active body but of the stilled body to be developed? Acknowledging the many limitations of mobile methods, is it possible or indeed necessary to develop a range of new methodological tools? It seems that perhaps rather than claiming the need for some epochal shift in approach (Crang, 2005), a certain degree of reflective criticism may be useful in adapting our current methodological toolbox to acknowledge and take greater account of these phenomena. For example, such experiences could be incorpo- rated into some of the more traditional approaches such as the semi- structured interview or even the focus group. Instead of focusing on the active agentive dimensions of experiences, questions could be recast with the aim of focusing attention on some of these more non-agentive sensibilities. It may be possible to talk *around* the experience of cor- poreal stilling in a focused but limited way such that quiescent experi- ences can be alluded to and verbally narrated (see Bissell, 2010). If the limitations of such textual accounts are appreciated, then these methods may provide a more comfortable and familiar entrance point for both the researcher and participant when confronting these phenomena.

Similarly, the emerging plurality of visual methods (see Rose, 2001) may also be well attuned at excavating some of these quiescent sens- ibilities which are perhaps not so readily narratable through texts. Whilst video-based methodologies are increasingly popular tools to record and narrate mobile experiences (Laurier, 2004), perhaps by returning to still- photographic images we can attempt to portray, albeit in a limited way, some of the lack of active, agentive movement emergent in certain situ- ations. Rather than acting as a productivist data-generative tool, these images could be considered and utilised as aesthetic products in their own rights (Crang, 2005). These photographic montages could chart the corporeal experiences of stilling in a dialogue with the reader, perhaps presented in series such as the stills used by Laurier and Philo (2006). Furthermore, such images could be adapted artistically to specifically draw out some of these quiescent sensibilities perhaps through the use of different contrasts or tones. Such images have the capability to affect the viewer by transmitting and circulating a variety of sensibilities (see Lisle, 2009; Jacobsen-Hardy, 2002; Steiger, 2000). The power of such a medium is apparent in the evocative black and white photography of Eugene Atget's Paris which documents an exquisitely stilled and serene sense of passage through the urban landscape (Nisbit, 1992). Ross Harley's (2009) images of light boxes in airport spaces also brilliantly convey a sense of the still. Here, stills can act as a proxy for the actual experiences of corporeal stilling and quiescent experienced. Whilst these quiescent phenomena that we are interested in here are beyond the purely visual, felt in body tissue, in the stomach, in the head, such sensibilities occasionally reveal themselves in sites of affects in process such as the face (Thrift, 2004). However, unlike more traditional textual empiricisms, these experiences cannot simply be inferred from and read off from these images. Whilst images are performative in the way they transmit a series of affectual sensibilities to the viewer, it must be remembered that the moments of image construction and image viewing are two discrete events where different affectual sensibilities emerge (see Crang, 1997). The viewer cannot *feel* the sensation of corporeal lassitude of the stilled subject by casually looking at the images since those experiences are anterior to the ongoing experience of the event itself: of waiting; of travelling. In translation, such sensibilities are often lost. Again, the acknowledgment of the specific limitations of the method must be at the heart of any such approach.

Perhaps these visual and textual methods should be used in combination to provide a more kaleidoscopic narrative of these quiescent sensibilities. Multi-method approaches have long been heralded as a means by which researchers can investigate a phenomena or sensibility *more fully* since different methods get at different aspects of the same phenomena (Bryman, 2004). Suspicion clearly hangs over such a sentiment that more *and different* somehow provides a superior comprehension. Nevertheless, to use a range of different forms of knowledges, such as visual and verbal and perhaps tactile, together could provide an ideal multivocal conduit for excavating a range of quiescent sensibilities. Crang (2005) highlights the importance of considering how different methodological practices relate to each other. Specifically Crang argues that the visual and verbal pertain to different ways of knowing. This may be true, however the differentiation of present-ational forms, between vision and text, still doesn't specifically target these experiences of stilling and quiescence. Rather, they are illum-inated by their absence. Gaps...spaces...hiatus. The researcher, whilst perhaps adamant to somehow capture these sensations, reveals more about the power and potentiality of absence: the *absence of words*, where communication stutters, fails and language ruptures; and where the often all-too automatic lens of the camera fails to focus. Perhaps we should look at quiescent through these absences and silences? To pri-vilege silence is certainly nothing new since much postcolonial liter-ature focuses on how silences can be utilised as resistive practices (see Lambert, 2005; Domosh & Morin, 2003, for example). However the problem with this is that such silences are only made apparent through their radical relationality to that which is active. To research quiescent sensibilities through the lens of silences or absences still serves to privilege active, agentive and auto-affective sensibilities, or rather view their absence as *failed action*, where action for some reason *did-not-occur*. Instead I want to chart an empiricism that treats quiescent sensibilities not as an adjunct to action or indeed even stasis, but as a sensibility that transcends this dialectic and as a sensibility in its own right.

Sensibilities of quiescence clearly do not lend themselves easily to re-narration. Whilst they are an integral quotidian experience, they are singularity in that they are unrepeatable; unknowable. So how do we get at a sense of this transcendent sensibility without resorting to tracing or hinting at absences through passages of text or images? Paradoxically how do we narrate the 'unsayable, the outside of lan-guage...the non-conceptual?' (Crang, 2005: 231). Perhaps the key to this predicament is in the style of narration itself. Our current pre-sentational grammar is arguably ill at ease and not sensitive enough to deal with the complexity of subjectivity. If we are restricted to textual media through our presentational outputs, then it would seem imper-ative to develop new narrative styles and presentational grammars which are more attuned to attend to these sensibilities. Such a response could possibly emerge through more experimental poetic styles of nar-rative which move beyond the stylised idioms of typical academic prose in favour of more creative, incomplete, wandering accounts. Of course experimental narration has a long history through the writings of Kafka and Proust and perhaps more recently through contemporary travel writers such as Diski (2006) for example. However, such wandering styles could be increasingly developed by human geographers to attempt to get at these elusive quiescent experiences. This may be a tall order, particularly since some attempts by human geographers at weaving more poetic narratives have often descended into the cringeworthy and sometimes downright embarrassing, perhaps illuminating more about our lack of narrative fidelity than the phenomena being described (Brunt, 1999; Sanders, 1999; Gans, 1999). Nevertheless to develop different narrative forms is surely an admirable aspiration and therefore our practices of narration could be reshaped not with the aim of producing some grand linear narrative but more modestly to provide splinters of sensibilities as illustrated through Georg Perec's '*Je me souviens*' (Becker, 2001) which consists of a series of short numbered paragraphs with the aim of illustrating fragments of modern life.

It seems to me that the goals of such an empiricism need to be rather more modest. Rather than developing methods which aim to reproduce some implied totality or fullness of experience, it might be more fruitful to take a more fragmentary, non-linear approach to narrative. This style of narrative might be more responsive to the complex, subtle and shifting nature of a subject that is *not always* engaged in auto-affective action. Such discontinuous methods of narration must respect and attend to the gaseous nature of subjectivity that shifts into and out of focus: subjectivity as 'sometimes wandering and sometimes reassembled' (Merleau-Ponty, 1968 in Wylie, 2002: 446). To this end, Wylie's (2006) recent methodological intervention goes some way to narrating these shifts through a combination of text and photo montage of a walk along the South West Coast Path. The fluid relationship between fractured body and landscape and forms of absence are cleverly narrated through the use of text and photo montage. Responding to this shifting subjectivity, such a method is more attentive to the ways in which the body can slide between activity and quiescence.

Perhaps we are impotent to narrate quiescence *per se*. It may be that we cannot document these quiescent sensibilities but should rather focus on narrating these slides *between* presence and absence; between activity and stilling; between different forms of subjectivity. Movement, activity and auto-agency still seem to haunt these narrative reconfigurations. Most importantly, activity seems to be a pervasive residual in all of these attempts at narrating quiescence: spectral, haunting. A recognition of these enduring active traces reveals something more fundamental about the nature of quiescent sensibilities. Specifically it shows how quiescence as corporeal disengagement seems to be both highly fragile on one hand and extremely durable on the other. The ongoing temporality of practical empiricism necessitates that much of this narration relies on memory; recollections of these events. When experiencing such quiescent sensibilities as tiredness, fatigue or lethargy, since these sensations hold onto and descend through the body without auto-affective agency they could be characterised as extremely durable sensations, stubbornly difficult to shake off. However the work of memory reveals how fragile such sensations are. Far from persistent, these sensibilities act like bubbles, float away, or burst leaving only action-filled recollections. If quiescent sensations reside in body memory they may therefore only be recalled through the repetition of quiescent events (Mallot, 2006); through the re-enactment of tiredness, fatigue, hunger. If memory privileges action, activity and movement, then these fragile quiescent sensibilities may have already escaped prior to attempted narration in diaries, autoethnographies, interviews and focus groups. It is these memory traces that need to be held onto through this more modest empiricism.

Conclusion

Relatively passive experiences such as lethargy, tiredness, hunger and pain – experiences that transcend the dualities of speed and slowness; mobility and immobility – are some of the most inescapable and perhaps even inevitable aspects of life on the move. Yet they seem rather more difficult to pin down methodologically than their more active and practical counterparts that possibly lend themselves to narration more easily. Whilst a range of mobile methods have been developed which aim to get at the experiences of bodies on the move, I have suggested that they tend to privilege the more active dimensions of corporeal experience, drawing on presentational vocabulary that narrates the body-in-action. Whilst I have tried not to caricature what mobile methodologies are or might do – since their epistemological and ontological remits are diverse – methodologies that set out to narrate the experiences of bodies on the move have tended to capture only a limited dimension of what it is to experience movement.

In response to this, I have attempted to suggest some of the ways in which these more passive, quiescent experiences can be incorporated into our mobile methods. However in doing so, this chapter has revealed how fragile and illusive such sensibilities are when translated into narrative form; often apparent more by their absence than their presence.

Theorising about such phenomena might be rather more straight-forward than actually setting out to explore it empirically. Yet these experiences of corporeal passivity are *not* periods of absence that are devoid of meaning. But neither are they durations that should be interpreted by filling them with overactive text and meaning. Therefore narrative practices must be developed that are more responsive to these other, rather more refractory registers of mobile subjectivity.

Parallel to the acceleration of bodies and objects in an increasingly speeded-up world, academics following these mobile bodies have similarly accelerated not only their theoretical standpoints (see Gane, 2006) but most importantly their methodological practices. This is hardly surprising in light of recent peformative calls to 'make methods dance a little' (Latham, 2003: 2000). In order for researchers to get closer to the experiences of corporeal acquiescence perhaps it is ultimately our style of research that needs to be *less hasty*. This is clearly not an easy task, particularly in an increasingly output-driven research climate where turnaround times of research products together with the limited range of recognised research outlets arguably stifle creative methodological development, perhaps suppressing the development of slowed methods. Despite these caveats, rather than movement and speed as vectors *for* empirical investigation, it may be more useful to consider the speed of the research itself. Indeed Hinchliffe and Whatmore (2006) have similarly called for presentational styles that are responsive to the unfinished complexity of matter unfolding which requires a style of research practice that is more *slow-paced*. In order to appreciate the non-agentive passive sensibilities of bodies on the move, we as researchers might need to be more modest in our goals. Only through the adoption of such fractured, shifting and critical styles of empiricism, will a more responsive narration of subjectivity emerge: responsive not only to the vast range of corporeal sensibilities but also to the limitations of empiricism.

Part II
Steering the Mobile

5
Liverpool Musicscapes: Music Performance, Movement and the Built Urban Environment

Brett Lashua and Sara Cohen

Introduction

> 'I actually never played there. I was due to, and then I moved away to London and when I came back, it was gone.' – Chris, a Liverpool musician, whilst walking past the former venue The Picket

Music has been closely related to movement. Popular musicianship, for example, is commonly spoken and written about using metaphors of mobility: musicians go out on the road, on tour, or gig in the club circuit (Laing, 2008) and most, like the musician noted above, come back. The language of popular music is suffused with movement. For example, the musical and stylistic development of the Liverpool post-punk band Echo & the Bunnymen was described by Reynolds (2005: 439) as a voyage:

> If Goth took one route from post-punk back to loud-and-proud rock, Echo & the Bunnymen followed another path: not descent into darkness but soaring into the light. The celestial drive of their crystal guitars and the seeking beseeching vocals conjured a sense of quest for a vague grail or glory. The Bunnymen pioneered a style of purified eighties rock.

At the same time music, particularly popular music, has also been described in terms of the urban, as illustrated by books on popular music entitled 'The Sound of the City' (Gillett, 1983) and 'Urban Rhythms' (Chambers, 1985). For Krims, music is 'specifically a manner of experiencing urban geography' and 'constitutes both a voyage through urban geography and a musical voyage' (2007: xv, xvi).

This chapter builds upon this dual emphasis on movement and the urban, examining relationships between popular music and the built urban environment by focusing on the voyages of musicians through urban space. Whilst Reynolds traces a typical story of musical develop- ment, the chapter considers the everyday musical routes, paths and voyages of musicians and the city spaces they engage with along the way (Homan, 2003). Such movements are important, not only because they spotlight the varied and (often) obscured trajectories of popular musician- ship, but also because by focusing on them we may inquire and discover more about the intricate webs and shifting patterns of urban life. On the one hand, therefore, we are interested in the routine movements of musi- cians around and across the city. On the other hand, we are interested in what might initially appear to be the antithesis of movement: the built environment, incorporating spaces and places for music-making, whether live music venues, clubs, pubs, recording studios, rehearsal rooms, cafes or domestic homes. Yet, as Doreen Massey has noted, these sites are not static entities, but are instead comprised of an 'intense multiplicity of tra- jectories' (2000: 226). That is, music venues and sites (and the built urban environments they in part constitute) have their own trajectories and his- tories, their own kinds of movements and changes which intersect with musicians' trajectories and histories in particular ways.

The chapter draws upon ethnographic research we have conducted on voyages of musicians in and around the city of Liverpool, and focuses on events related to live music performance venues in a small area of Liverpool city centre known as 'the RopeWalks'. First, we focus on the minutiae of the soundcheck to show how music performance involves regular voyages that play a central role in the process of becoming a musician and in the creation of musical spaces, places and landscapes. Secondly, we illustrate the voyages of musicians in Liverpool through a discussion of maps and ethnographic mapping, focusing on stories musicians tell about their music-making, and featuring hand-drawn maps musicians have made to represent their movements around the city. The third part of the paper focuses on a different kind of musician's voyage – a walking tour interview, or ethnographic 'go- along' (Kusenbach, 2003) – to illustrate how research that not only focuses on movement but also incorporates it as a methodological tool can help to further the study of the relationship between music and the built urban environment.

Movement, musicians and urban environments: The soundcheck

The soundcheck is generally comprised of several hours before a live music performance when musical equipment is set up, microphones are placed, and the overall volume and mix of the performers' voices and instruments are set in the venue sound system. Soundcheck is mundane work, part of the routine, commonplace practices of musicians. These activities are easily overlooked or ignored for their tedium and apparent triviality, yet soundcheck is important not only to the sound of performance but also in terms of the movements and relations of musicians in urban environments. As part of our research Brett has been attending not only soundchecks but also musical performances, rehearsals, studio recording sessions; accompanying bands to gigs, helping bands with gigs (as road crew or 'roadie'), and also performing as a drummer (the researcher as musician *manqué*); and generally engaging in the kinds of 'hanging out' that comprise much of the time of popular musicians. We have been participating in and observing music-making activities in order to trace the journeys and everyday routes of musicians, documenting musicians' use and experience of the urban environment, mapping the physical and geographical spaces they inhabit and transform for music-making purposes, and considering how their music-making might influence and be influenced by, the appearance, sound, feel and atmosphere of the urban environment. This kind of participant observation is crucial if we are to better understand what it means to be a musician and the relationship between music-making and urban environments. As Cohen (1993: 124) explained, the ethnographic researcher aims to learn the culture or subculture they are studying and come to interpret or experience it in the same way that those involved in that culture do, that is, to discover the way in which their social world or reality is constructed, and how particular events acquire meaning for them in particular situations.

In terms of live music performance, the soundcheck has proved to be a time of privileged access for Brett as a researcher to see what is going on up close; during the rest of the night he generally won't be allowed on the stage but can watch from the floor with the rest of the audience. If the geography of a gig is constructed in crude halves, the stage (and backstage) is largely sacrosanct – even in small venues but especially so in large halls and arenas – delimited as out-of-bounds for audience members. Performers may, of course, mingle with the audience if they choose. As a researcher, however, Brett's movements through many venues could be considered to be liminal in the sense that they were affected by the classic ethnographic tension between insider and outsider status (Bennett, 2002). Bennett argues that more than merely acknowledging this tension, ethnographers must also consider (following Marcus 1998: 190), how reflexivity informs the researcher's position. In this case, we considered how critical insights into gigs were derived from Brett's status as more than a fan but less than a performer. Brett had some privileged access to the musicians he had interviewed and was sometimes able to mix backstage, but he wasn't a fully-recognised insider. Alternately, as a member of the audience, standing amidst the crowd during a performance, he had insider knowledge of the goings-on backstage and on stage. Yet – unless helping out with road crew duties or invited to hang out – like the majority of people at the gig he often wasn't allowed to pass beyond certain doors and gatekeepers into backstage worlds. Thus, during a live performance, the power geometries (Massey, 1993) of musicians' and audiences' mobilities could not be much clearer. The relations between spaces in front of and behind the stage are both produced by, and productive of, the (im)mobilities of the musicians and audience. However, these movements during the live performance are but a fraction of musicians' day-to-day activities, relations, and experiences. What of their mobilities at other times, such as during the soundcheck, when the boundaries between spaces and roles are much more fluid? And what then of the roles – and mobilities – of the researcher?

Korova is a small, trendy bar with a basement live music performance area, and it is located in the RopeWalks district of Liverpool. Formerly a locus of the city's rope-making industries, the district's long, narrow streets reflect the space required for the manufacture of rope used to rig the sailing ships that once frequented this port city. The rows of warehouses which stood derelict for decades have been transformed, since the 1990s, into a distinct entertainment quarter, home to many live music venues and dance clubs (Cohen, 2007; Gilmore, 2004). In the early evening, Korova's basement is nearly empty and very dark. A few floodlights spill over the stage, and the handful of people present pass spectrally in and out of the darkness, scrambling about their business, which is the mundane work of setting up amplifiers, connecting cables to microphones, and assembling the drum kit. The volume of each individual instrument, drum and microphone must be set through the house PA and on-stage monitoring systems – this is the soundcheck. There are a thousand small tasks to attend to, from tuning guitars to taping down loose wires, to setting out bottles of water for later thirsts. Everything must be in its right place, and we scuttle about the stage to get things set. The drummer starts hammering away on his kit to ensure nothing is likely to fall over, the bass player turns on his amp and begins picking through a riff, while across the stage the guitarist opens up with some thunderous chords while shouting 'check, check, check' into his microphone. After a few moments of this noise, they pause. The drummer then counts off '1-2-3-4!' and the band launches effortlessly into a song together. Of course, an incredible amount of effort went into producing this moment, which is only a soundcheck to an otherwise empty room, but the moment is nonetheless a remarkable convergence of people, objects, and labours.

Prior to the soundcheck, we had met to collect the assorted musical equipment (drums, amplifiers, guitars) from a rented rehearsal space in a dank, musty, and cluttered basement rehearsal room. This red-brick Victorian building half-way across the city is where the band practises one or two nights a week. We hastily shoved the band's gear into a small, battered, untidy car – littered with fast food and sugary drinks containers, the cheap fuel of many musicians who eat on the go – and similarly unloaded it again at the venue. This process would be repeated in reverse (and with even less care) in the small hours of the morning, packing up the musical equipment from the venue, loading it into the car, driving it back to the rehearsal rooms and carrying it back down into the basements.

The soundcheck draws toward completion and after the venue's doors open there can be no more of this clamorous racket. The mantra of the soundcheck is: hurry up and then wait. It is approaching 7.00pm and we must finish soon, even though the band won't appear on stage, on this night, until after 10.00pm. Ironically, the live public performance – the time when supposedly all of the 'action' happens – is perhaps the most static moment for the equipment and the musicians involved in the production of the evening: an hour in a fixed location, guitars not carried or carted here and there, amps not moved on or offstage, drums not packed or unpacked, nothing driven around in cars or unloaded from vans, nor plugged in or unplugged. Nevertheless, the musicians do move 'live' on stage. These movements – even when appearing most brash and unrehearsed – are the products of much practising and repetition.

In a few days, or perhaps a week's time, many of these events will be repeated, with minor variations, by the same musicians at another music venue elsewhere in the city. As Thornton (1995: 91) has noted, 'the seemingly chaotic paths along which people move through the city are really remarkably routine.' While live music venues differ in size and location from small rooms for hire, to local pubs and bars, and from concert halls all the way up to massive arenas (pictured), the rituals, customs and voyages of a soundcheck are largely the same. These kinds of voyages are central to the processes of becoming a

Fig. 5.1 Graham Jones of the Liverpool band Voo during the soundcheck at the Echo Arena, July 2008.
Source: Courtesy of Pete Martin

musician. This is suggested by Ruth Finnegan's (1989) use of the term musical 'pathways' to describe the everyday routes and routines and flow and flux of music-making in urban life. These musical pathways are forged through the activity, hard work and commitments of the musicians involved. They are known and regular routes through life, and just some of the many pathways that people take in their lives. They are also the regular spatial routes that musicians take around, across and in and out of the city in order to participate in music events and activities – including the soundcheck.

Building upon the work of Appadurai (1996) on the 'production of locality' in everyday life, it could be argued that these routes are not just musical pathways within urban space but part of the process through which urban space is produced. Through their music-making musicians inhabit, move through and interpret urban environments, defining and distinguishing musical spaces and places, marking out the boundaries of who belongs where and when and thus 'territorialising' social space (Deleuze & Guattari, 1987). Duffy (2000) illustrated this kind of territorialisation in her study of music festivals in northern Australia. She described, 'Deleuze and Guattari's (1987: 316) understanding of the relationship between place and performativity as one of territorial possession, where a territory is marked out through artistic practice and where possession is made known through being marked by this art' (2000: 59). These artistic practices enabled musicians 'to attach and reattach themselves to a sense of place and make claims of belonging' (Duffy, 2000: 63). Our example of the soundcheck in Liverpool illustrates similar processes by highlighting the territories and boundaries that it produces, and the mobilities of the musicians involved as well as those of the researcher.

Representing musical mobility: Musician's maps and mappings

In addition to accompanying musicians to gigs and soundchecks, as part of our research on relations between musicians and the built environment we have been asking musicians to sit down in an interview with us and produce their own hand-drawn maps showing sites and journeys connected to their music-making, as well as drawing into local maps that we have provided. In this regard, maps – for all of their propensities to immobilise and deaden the movements and flows of everyday life (Chambers, 1985; de Certeau, 1984; Massey, 2000) – have proven, nevertheless, to be an effective tool for prompting narratives and memories of musicians' mobile practices. They have also shown dramatically how the city and its musical landscapes have changed over the years. The anthropologist Marc Augé wrote in his ethnography of the Paris subway system, *In the Metro*, that 'it is a Parisian privilege to use the subway map as a reminder, a memory machine, or a pocket mirror on which some-times are reflected – and lost in a flash – the skylarks of the past' (2002: 4). With such memory machines, he continues, 'travellers suddenly discover that their inner geology and subterranean geography of the capital city meet at certain points, where dazzling discoveries of coincidences promote recall of tiny and intricate tremors in the sedimentary layers of their memory'. We have been using maps to prompt musicians' stories of the relations between their music-making practices and the built urban environment. As Turchi noted (2004: 11), 'to ask for a map is to say "tell me a story"'. Such stories provide useful commentary about how changes to the city have shaped musicians' pathways, and in turn how such pathways may (re)shape the physical fabric of the city.

For example, here is a map drawn by a Liverpool singer/songwriter who has been actively performing in the city for over 20 years. Such maps echo the hand-drawn maps of passengers on the London Underground

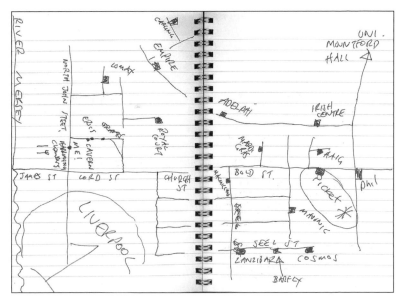

Fig. 5.2   A Liverpool singer/songwriter's hand-drawn map

described by the artist Helen Scalway (in Pile & Thrift, 2000). These sketches – which Scalway alternately refers to as drawings, maps, and pictograms – provide tracings of the city by which traveller's 'hands re-enacted their different experiences of space' (2000: xvii). The sketched maps evoked travellers' movements within an 'unlocatable territory between actual routes, imagined routes, and the routes officially mapped' (2000: xvii). On the musician's hand-drawn map (above), music venues emerge as nodal points at the intersections of his musical pathways and the changing built environment of the city; the map provides a partial diagram of memory and experience and the development of his career, as well as showing how the city has changed over time.

First, this map may be read almost as a chronicle of legendary rock performance venues in Liverpool from the 1980s and 1990s, many of which have since closed, moved, or disappeared completely, including Eric's, the Lomax, Cosmos, the Irish Centre, and the Picket. While sketching his map, the singer/songwriter spoke of MacMillans, a venue located in the RopeWalks district that marked the start of his musical career, and the Cosmos – in the 1990s getting a gig there meant you were on the right track to success. He related further narratives of the gig when he knew he'd arrived as a professional musician, at the Royal

Court venue, and other venues in Liverpool where his performances signified important moments in the development of his career, such as playing a concert at 'the Phil' – Liverpool's Philharmonic Hall.

Second, this musician highlighted in particular the venue called the Picket. When drawing this map he had initially forgotten to draw in the Picket as a live performance venue. This may be connected to the fact that the Picket was forced to move in 2004 following the sale of the building where it was housed. The Picket had opened in 1984 in the Merseyside Trade Union Community and Unemployed Resource Centre on Hardman Street. It has since relocated to a newly developed 'Independent arts cultural quarter' launched partly to house and promote music and arts activity displaced from other cultural quarters nearby as a result of commercial development and gentrification. Thus certain venues signify particular areas of, and changes to, the city and help mark out city spaces: For example, the Cavern Club and the Cavern Quarter[1] symbolise, perhaps, the apex of local regeneration based on cultural and heritage tourism. Other venues, such as the Picket, have sought to create links (symbolically, politically, socially, and economically) with the city's working-classes: the Picket was housed in an unemployment centre and initially had been set up by, and for, unemployed young musicians. It now hosts a working class music festival and serves as a key venue in an area envisioned by local organisations as a kind of counterweight to the heavily commercialised 'cultural quarters' in the city centre. Mapping the movements of musicians to and from these and other venues has brought these kinds of characterisations and changes to the city into sharper relief.

Thus we have found maps to be a useful technique for tracing the journeys of musicians and evoking narratives of the changing city. Through the verbal story-telling of musicians, venues emerge as 'articulated moments in networks of social relations and understandings' (Massey, 1993: 66). These kinds of mappings – either hand-drawn or traced on printed maps – provided representations of the routes of musicians through the city. That is, they provided one way of showing the importance of movement to musicians' practices, as well as how musicians remember and re-imagine the city. Furthermore, in order to gain a better understanding of these mappings and 'articulated moments', we found it useful to *move* with musicians, to fall in with the repetitions and social rhythms of musicians' ordinary comings and goings – from attending soundchecks to rehearsals in old, musty warehouses, from having a pint in a local pub to sessions in recording studios, and from cafes to live performance venues and festival events. Some of these movements with musicians, related to moving through and remembering the city, we consider below in the third and final part of the chapter.

Movement, memory and urban change: A walking tour

In addition to drawing maps for us, some musicians also agreed to take us out on walking tours of the city, showing us around and (re)tracing their musical paths through the city. A walking tour interview, or 'go-along' (Kusenbach, 2003), refers to this deliberate practice, somewhere midway between, and combining the virtues of, the sit-down interview and simply 'hanging out'. Kusenbach suggests that ethnographic enquiry set in motion along 'natural' and ordinary itineraries can sensitise researchers to 'the substantial role of place in everyday social reality' (2003: 477). For us, walking tours are useful in several regards but first and foremost because the cityscape itself often acts as a prompt or 'memory machine' as we move through the urban environment.

This section of the chapter shares an example of a walking tour conducted with the singer of a Liverpool group that had several top-ten hits in the UK during the late 1980s and early 1990s, and a remix (of an earlier single) that also charted in the top-ten in the mid-2000s. During Liverpool's tenure as European Capital of Culture 2008, the band was enjoying renewed attention, with several head-lining performances in the city as well as a national tour. The lead singer and Brett first met at a cafe for a semi-structured, informal interview. During this conversation, rather than draw a map, this musician offered to take Brett around the city to show him the locations of key music performance venues that had played an impor-tant part in the story of his life as a musician but were no longer 'on the map' – having been converted since into flats or trendy night-clubs, or in some cases demolished altogether. The walking tour in effect re-mapped the city via the musicscapes of this singer's memories. Instead of a map, the built environment often prompted memories of former routes, sites and performances. Indeed, the route the walking tour took often meandered and detoured as once-forgotten venues and former sites of musical activities sprang into the musician's memory, cued by the built environment as we moved through it.

For example, the singer took Brett through the RopeWalks district towards the site of a venue once known during the 1980s as The Warehouse. They walked across an open square bordered by old warehouses that had been converted into nightclubs and restaurants. There the singer shared a short narrative of his early career, interwoven with the story of a particular building that borders the plaza:

So we're walking through Concert Square now, which back in those days, there was nothing here. These would all have been working warehouses [once]...in the late eighties, this building [*pointing*] was the start of bohemian Liverpool in many respects. This, where [current nightclub] is, this was all rehearsal spaces, cafes, and independent organizations. And there was an old chap who used to rent it out and we just heard from our drummer that there was rehearsal space here and they had a music project based in there and this was the front door to it. [*Pointing*] That was the café there. The next part of this development of this area was just a bit further up where they had 3 Beat Records started up, about '89, '90. This was the main entrance to the café, can't remember the café's name. [The drummer] might remember, but the rehearsal space was in here. You had lots of bands playing in here and then we knew something was happening 'cos in early 1990, James Barton [later of Cream] was making a name for himself as a bit of a DJ and he came on a bit of a tour with us, and he said we're opening this new place and it was this complex here...We opened a café downstairs in this building and there was a café and it was all independent shops but it was the first thing in this area and that was definitely 1990, yeah we opened up a café downstairs...Only a small venue but we played a few songs and it was like people queuing outside the street, you know it was like you could tell something was happening, it was a bit of a buzz in the area. All that space there [*pointing*] these were the load-ins for the rehearsal rooms and it didn't have a name as such, Holmes Buildings I think it was. And [the band's] office was in here. Yeah Holmes Buildings that's what it was called. So all these big giant doors here, these were the load-ins for the rehearsal rooms. It was rehearsal space or artists, you know, and music projects, but mainly groups rehearsing, that was the main activity in that building. You had photographers as well and graphic designers. You wouldn't have a clue that that type of thing had gone on now, you know, looking at it. As you can see it's an old warehouse. Used to have stuff from the docks coming up, I don't know what they used to...It had a terrible smell of like, you know, strong smell of things that had been stored there over the years.It wasn't a bad smell; it was just

Fig. 5.3   Concert Square, RopeWalks, Liverpool, 2008.
Source: Courtesy of Alice Lashua

a strong smell. I'm pretty sure the van used to pull up here and we'd load in.

Strolling through this area of the city thus provided another example of an important 'discovery of coincidence' (Augé, 2002) triggered by walking through Concert Square. In an earlier sit-down interview, this musician hadn't mentioned this space or its importance to either his musical career or the varied regeneration initiatives in the city. A key point we make here is that ethnography can provide a useful approach to the study of relationships between music and the built urban environment, particularly ethnographic approaches designed to not only observe and record movement but to incorporate movement as a methodological tool and a means of participation. The walking tour initially set out to tell a story of one person's musical career in the city by revealing the spaces and places that have played a key part in this story, but the to-ing and fro-ing of the touring interview transformed it into a story not only about music and the built environment but also about urban change. According to this musician, the regeneration of the RopeWalks began with the Holmes Building in 1990. Thus we can link this narrative of memory, prompted by the walking tour, to the transformation of an entire district of the city centre, and the near-mythic origins of the creation of a 'cultural quarter' around arts, creative industries, and nightlife.

Conclusion

The musician cited at the start of this chapter noted that he never had a chance to play at the Picket because the venue had closed and relocated while he was away, living in London. Much like musicians and their music, cities and built urban environments are never fixed but continually on the move. As Doreen Massey (2000: 227) wrote of her regular commute home to London from her workplace in Milton Keynes:

> This is not a matter of crossing space to a static place that has been somehow lying there, waiting for our arrival. You have to catch up with what's been happening, with how this place, too, has been moving.

Which is simply to say that Liverpool (or, as Massey might say, the multiplicity of trajectories that is called Liverpool) is constantly changing, shifting, and moving, and music-making is both productive, and also a product, of the city: 'space produces as space is produced' (Leyshon, Matless and Revill, 1998: 425).

This chapter has stressed the importance of musicians' movements and voyages in processes of spatial production. In doing so, we have not discussed perhaps more obvious examples, such as musicians on tour, or 'on the road', or turned to analyses of lyrics or songs. Musicians' tours are in themselves fascinating case studies and these mobile practices have often played a significant part in popular music mythologies. They have been represented in films[2] about popular music, for example, and also in books.[3] Lyrics often intone the movements and routes of musicians too, as well as provide links to memories of the changing city. Here we would be remiss not to mention songs such as The Beatles' 'Penny Lane' and 'Strawberry Fields Forever', and their connection to built environments, but also their mythic reconstructions of a suburban pastoral (Daniels, 2006).

Here, however, we have focused on the more routine, ordinary, and everyday movements of musicians in order to explore relations between popular music, the built environment and urban change. Using one particular event – a soundcheck – as a starting point, the first part of the chapter discussed the mundane journeys taken by musicians for the purposes of music-making. These activities are central to the processes of becoming 'a musician' and the creation of musical 'pathways' (Finnegan, 1989). Our research approach, spending in-depth time in close proximity with musicians and also moving them, is in sync with approaches that view ethnography as more than a method, and instead as a process of creating and producing knowledge – through experiencing, interpreting, and representing cultures and societies (Pink, 2007). In the second part of the chapter we then discussed the representation of movements and journeys through a musician's hand-drawn map and accompanying narratives. The third and final part of the chapter focused on a musical journey that enabled the researcher to move along with one of the musicians participating in the research. This walking interview drew attention to the way that the built environment can prompt memories of the musical past, but also to changes and movements within the built environment and their impact on local music-making. Our focus on movement has thus helped to illustrate how popular music-making characterises cities and material urban environments, and how in turn the transformation of cities and urban environments affects popular music-making. The power of ethnographic studies of musicians' mobilities, for our research, resides in the ability to observe and question everyday, ordinary and habitual musical practices, and connect the routine comings and goings of musicians to broader theoretical contexts in relation to the production of urban spaces.

Notes

1   The Cavern Club, made famous by the Beatles (or, perhaps, which made the Beatles famous), was demolished in 1973 to make way for a ventilation shaft (that was never built). In 1984, a replica Cavern Club was opened in a cellar two doors from its original location, with the 'Cavern Walks' shopping centre above it. The venue, and surrounding area, has since become a buzzing nexus of bars, clubs, shops, tourism and heritage sites (see Cohen, 2007: 155–183).
2   For example, the films *Almost Famous* (2000) and *That Thing You Do* (1996) relate stories of fictional music groups through the narrative device of being 'on the road'. The band's journey (literally) drives the plot – a sudden assent to success and fame, conflict and internal struggles due to the changes that fame brings, dissolution of the band, and some sort of redemption. The tour is central to the construction of the story, as well as the construction of a particular mythology of popular musicianship.
3   Bolland's (2006) chapter on Liverpool musicians of the 1970s is titled 'Vans, beer and missing bags', and another example is Trezise's reportage of the Welsh band Midasuno on tour in *Dial M for Merthyr* (2007).

6
Vim de Bahia pra lhe ver: Multiple Movements in the *Capoeira Batizado*

Neil Stephens and Sara Delamont

The title, which translated from Portuguese means 'I came from Bahia to see you', is the opening line of a well-known song used in the Brazilian dance, fight and game *capoeira*. It can be sung in a cheerful, friendly way, as a simple greeting, or with menace, carrying the meaning that the singer has travelled to engage in a serious contest with the listener. Other places can be substituted for Bahia, so a *capoeira* master from Sydney visiting Auckland could sing 'Vim de Australia pra lhe ver' to produce smiles among the New Zealanders. We have heard and sung the song many times. The paper focuses on the polyvalent meanings of one *capoeira* kick (*the armada*) which, like the song, and all *capoeira* movements, can be menacing, welcoming or carry other meanings.

'We', the authors, two sociologists, are *Trovao*, a *capoeira* practitioner and *Bruxa* a non-participant observer. UK *capoeira* teachers give students a nickname when they join a group. These Portuguese names, meaning Thunder and Witch, are our pseudonymous *capoeira* nicknames.[1] We have written about our research methods (Stephens & Delamont, 2006a) and other aspects of diasporic *capoeira* in the UK elsewhere (Stephens & Delamont, 2006b, 2008, 2010a, 2010b, 2010c; Campos de Rosario, Stephens & Delamont, 2009) and have minimised duplication here. Fuller bibliographies of the literature on *capoeira* can be found in those papers. Here we draw on five years of joint ethnographic fieldwork to reflect on multiple forms of movement, focusing on the periodic festivals, called *Batizados* because they include the baptism (*batizado*) of novices into *capoeira*.[2]

Capoeira is a dance and martial art done to music, with origins in the African-Brazilian (*i.e* slave) population of Brazil, which is now widely taught around the world (Assuncao, 2005).[3] UK students enrol with a teacher (usually Brazilian), take classes two or three times a week, and enjoy festivals of their own group twice a year, as well as visit other clubs for *theirs*. Most *Capoeira Regional* groups have a system of belts (*cordas*) that show the learners' level of achievement.[4] Early *cordas* are given for regular attendance and minimal competence at the physical and musical skills (all students have to clap and sing during *capoeira* play and be learning the five musical instruments), and come at yearly intervals. More advanced belts require much greater competence and commitment, and are spaced at longer and longer intervals. *Trovao* got the first three belts in 2003, 2004 and 2005, but has not yet been scheduled for the fourth. Festivals are fun for all students, whether getting a *corda* or not, because there are classes by visiting, sometimes famous teachers bringing new skills, parties, public performances, reunions as absent members return, and new things to buy.

Movement and *Batizados*

Many kinds of movement, real and symbolic, occur at, and around, the *capoeira* festival. Consequently these events are an ideal focus for innovatory research methods that do not freeze mobile experiences. Drawing upon Sheller and Urry's (2006) 'new mobilities paradigm', focusing on mobile practices rather than social processes, we follow Urry (2007) in proposing mobile research on mobile bodies. Many types of mobilities are found at *Batizados* and after outlining five of these, we focus on the polyvalence of one kick. Only mobile methods can unravel the multiple meanings of that kick or any other 'move' in *capoeira*. The five mobilities are:

1. Physical travel. *Capoeira* festivals attract travellers: teachers come across the world, some from Brazil itself, students come from all over the country, and from neighbouring countries (e.g. to the UK from France, Germany and Ireland). The festival of a popular teacher can attract twenty or more teachers and over 100 'guest' students.
2. Physical Movement. All these people have travelled to move. Several days are spent doing physically energetic *capoeira* and Brazilian dance, and just as important, dancing all night at parties.
3. Movement across boundaries. The concentration of Brazilian teachers, and the meeting of people from many different groups, mean there is a great deal of clustering, boundary maintenance, and boundary crossing, visible at festivals.
4. Symbolic movements of several types occur. Students from the host group who are being baptised move symbolically *into* a *capoeira*

community. Others, whose teacher has judged them to be ready, are promoted to the next belt in the hierarchy. These are classic *rites de passage*.
5. Researchers' Movements. Festivals are occasions to study all the previous types of movement, physical and symbolic, in a concentrated space and period of time, intensifying all aspects of the research.

We have illustrated all five types of movement, by focusing on the multiple meanings that one *capoeira* kick – the *armada* – can have at a *batizado*. We expand a little on the five types of movement, before using the armada to reflect on the most sociologically interesting types. After the examination of the polyvalent meanings one movement can have we draw out the more general merits of such readings of movements, by mobile researchers, at the end of the paper.

Movement through space

The *batizado* is predicted on movement. Novices are given their *cordas* by visiting teachers, who welcome them into *capoeira* or promote them to a higher grade. Consequently every teacher has to invite guest teachers to his *batizado*, pay their expenses, arrange accommodation and meals for them, and show public respect to them. In return, he has to go to other people's festivals, making arrangements for his own classes to be 'covered' in his absence.[5] A 'good' *Batizado* has visiting teachers *from Brazil*, especially a master (*mestre*) with a famous name or the UK-based teacher's *own mestre* or the head of the teacher's group. A famous *mestre* attracts more teachers and more students to an event. When leading figures in a group are scheduled to attend, teachers in that group travel across Europe with their students. In Spring 2008 the founder of *Cordão de Ouro* (Golden Cord) *Mestre* Suassuna was in London, and the *Cordão de Ouro* groups from Manchester, Birmingham, Nottingham, and Belgium travelled to London with their teachers to show respect.

Physical movement

There are many types of physical movement at *Batizados*. Festivals contain master classes in *capoeira*; lessons in African, or Brazilian dance; African-Brazilian cultural manifestations such as *maculélé* (a dance with clashed sticks) and the *puxada de rede* (miming pulling in a full fishing net); *capoeira* play in many *rodas* (a ring of people, clapping and singing, while two players do *capoeira*); parties with lots of dancing; and perhaps a public performance of all the above. These physical movements are highly valued. For those getting the next belt, the opportunity to play a *mestre*, however briefly, is the most important movement of the festival. For the teachers present, especially those who are *Professors* or *Contra-Mestres*,[6] or teach in cities without other *capoeira* groups, the chance to play *capoeiristas* of the same or higher levels is particularly attractive. After attendance at a festival away from Cloisterham, if we ask *Achilles* whether he had a good time, his first comment is usually 'I played lots of *capoeira*': that is, there were lots of opportunities for physical movement competing with players at his level.

Crossing boundaries

Festivals are characterised by many types of movement across many boundaries, and equally by the creation of boundaries, and the visibility of boundaries that impede movement. *Batizados* have spatial arrangements that encourage and restrict movements. The typical festival is held in a large hall: bigger than those used for normal classes. In one corner, there will be a table with food and drink for the teachers: water, juices, coffee, fruit, sweet things, savoury sandwiches and crisps. This food and drink is *only* for the teachers (and their wives and children if present), although a teacher might give food or drink to any child. Near the food is a space where the teachers congregate when not teaching, playing in the 'band' (*bateria*), or taking a class themselves as a student if the teacher is *very* distinguished. This space is Portuguese speaking, often noisy with lots of laughter, and symbolically 'out of bounds' to the students. Alongside the teachers' space is the 'shopping area', where they sell *capoeira* clothing, Brazilian clothing (e.g. bikinis, flip flops), DVDs and CDs of *capoeira*, instruments, jewellery, and sometimes books about *capoeira*. Students, especially young women, cluster in this area, hold garments up against themselves, and run backwards and forwards to the changing rooms to try them on.

Elsewhere in the hall there are chairs or benches for the *bateria*, where instruments are set out if it is not currently playing. Here teachers and more senior students mix: the musically competent students from the host club are expected to offer to play instruments to give the teachers a rest. Most of the students set up 'camps' along the remaining walls where they put bags and coats, food they have brought, water, and possibly their sleeping bags. Wherever there is a break, many go to 'their' stuff and sit down to eat, drink, change sweat-soaked T-shirts for dry ones, repair blistered feet or check text messages. During classes individuals run to their water bottle whenever they are thirsty. Groups of friends, or people from the same club, tend to cluster and if there is a meal break they sit and eat together.

The host teacher has a few reliable students as 'official' helpers, who move between the *bateria*, the teachers' corner (if needed) and frequently vanish from the hall to return with more food or drink, another teacher they have collected from the station, or further supplies of the 'local' T-shirts etc. These helpers often wear a different T-shirt so they are recognisable as locals. One or two people are 'on the door' taking the money, and tying ribbon round the wrists of the arrivals to signal that they have paid. Others may be handing out the event T-shirts, or meal tickets, or passes to ensure entry to the nightclub for the party. They may be too busy to take any classes. Sometimes they are 'injured' (i.e. those who because of sprained ankles or back pain are 'not training at the moment'). There may be someone videotaping the event for the organiser, and an 'official' photographer.

Boundaries can be linguistic or sartorial: the teachers are Portuguese speakers in a separate linguistic space. Students from a 'foreign' country speak their language (e.g. the club visiting from Utrecht speak Dutch) but students from different clubs who share a language may also cluster: at events we attend there are often students from Poland and other Poles gravitate to them. Clothing marks boundaries; students from separate clubs wear distinctive uniforms and people at various skill levels wear different coloured *cordas*. Some classes are for all abilities (especially dance lessons) and then students probably train near people from their club. Other classes are divided by ability, and then people are segregated by the level of their *corda*.

The crowds at *Batizados* are *not* normally segregated by sex or race. *Capoeira* is mixed, and men and women are found training together all over the room. The students are normally a mix of races and nationalities: typically when *Leontis* came to *Hercules*'s 2007 Burminster festival he had four students, two African-Caribbean men and two white women, one of them Spanish, with him. One of *Hercules*'s top students is a Japanese woman, the other two are white British men.

Researchers' movements

Our roles, and therefore our movements, are different. *Trovao* takes classes when 'away'. At Tolnbridge festivals he is usually an organiser, sorting out night club tickets or helping visitors find food, or beds while *Bruxa* tidies up rubbish, gives out T-shirts, or goes to the shops for water, fruit, plasters or even lavatory paper. Generally though, while *Trovao* is moving, doing *capoeira*, dance, and socialising, *Bruxa* is still:

writing fieldnotes in an unobtrusive and safe place during classes. In the gaps between classes, she is much more likely to cruise the shopping area than *Trovao*, and be escorted into the teachers' area by the host to be officially introduced to the visiting masters (*mestres*). At his 2008 festival, for example, *Hermes* took *Bruxa* into the teachers' corner to introduce her to *Mestre Suassuna* and *Mestre Gato*, the two most distinguished people present. At the end of a festival we are both tired, *Trovao* because he has been moving a great deal, *Bruxa* because she has been focusing intensely on the action and writing fieldnotes.

Symbolic movements

The most anthropologically interesting movements that occur at *Batizados* are socially symbolic: the *rites de passage*: analysed elsewhere (Stephens & Delamont, 2010c). However, any *capoeira* move can also have symbolic meanings, and here we focus on one: the *armada*.

The Armada: a polyvalent kick

We chose to focus on the *armada* because all *capoeiristas* use the same name for that kick (unlike some other moves that have various names), it is an easy name to learn and pronounce (unlike, for example, the *queida de rins*), it is taught to students starting *capoeira*, to more advanced students, and used by professionals. Many students will, in their first lesson or certainly their first few lessons, be taught *armada*. Meeting the *armada* is meeting *capoeira* itself. The *armada* is literally an armed strike, used in faster games when the players are upright, described in Capoeira (2003: 75–77), the best selling instructional book. The player turns her upper body partially away from her 'opponent', then spins her head through 180 degrees so she is watching him over her other shoulder, then moves her feet, to release one leg into a spinning kick that follows the head and the torso round. *Trovao* from his embodied experience of the movement stresses that (a) the kick is circular–with the foot making a 360 degree circle around the body – with knee straight; (b) that it is highly routinised and drilled to be identically performed by people (even though we show it has different meanings).

The *armada* involves moving all the main parts of the body and using the brain. In executing the kick the head, the arms and hands, the waist and torso plus the legs and feet, are all engaged. The player also uses learnt skills such as balance, having a good base, flexibility, aggression and maintaining eye contact with the opponent. The kick has to start from first position, the player has both feet on the ground, legs apart, knees slightly bent, well balanced (good base), watching her opponent. Then the feet move, the head swivels to keep eye contact, the arms go up to maintain balance and protect the head from a possible counter attack, the torso swivels, demonstrating the flexibility of the waist, and then the legs are used in the kick. One actually bears the body's weight and is vulnerable to being attacked by the opponent. The kicking leg is as high in the air as the player can get it when it is out straight, and is whipped round as fast as the player can rotate it. Delivered by a fit, skilled player the *armada* comes fast and with considerable force. Opponents escape it, usually by dropping underneath it. *Capoeira* attacks are not blocked, as in many other martial arts, but dodged. Escapes (*esquivas*) have names describing them, such as the *frontale* where the player drops to the floor facing the front, or the *laterale* where she drops to the side facing away from the kick.

The *armada* is polyvalent: like all *capoeira* moves (Almeida, 1986; Browning, 1995; Capoeira, 1995, 2002, 2006; Downey, 2005; Lewis, 1992). During a *batizado* all its many meanings are routinely visible. *Trovao* would perform it, *Bruxa* would observe it being done, in many ways, as we demonstrate below. Our description is drawn from many hours of participation and observation, interpreted through long conversations. *Trovao* has reflected on what every type of *armada* he does and sees means to him. *Bruxa* has asked teachers about their strategies. Both authors have discussed students' strategies at, and understandings of, classes and performances over a five year period of collaborative fieldwork (Stephens & Delamont, 2006a). We have drawn on the fieldwork and the conversations to produce a composite picture of the polyvalent meanings we recognise. More expert *capoeiristas* could 'see' more types and more nuanced meanings; novices would not 'notice' all the performative aspects of the kicks. Because we have been intensively engaged in learning about *capoeira* either as an embodied experience or with the ethnographic gaze we are relatively 'expert' observers of it. (See Stephens & Delamont, 2006a for more detail). For this paper we read all the fieldnotes from twenty *Batizados*, fixed on the eighteen manifestations of the *armada* that regularly appear at a festival, and then created an account that would be comprehensible to a reader. The eighteen varieties, and the interpretations were constructed for this chapter, drawing on at least 3000 hours experience of *capoeira* we have accumulated, especially the twenty *Batizados*.

A day of many armadas

On a typical festival day at least eighteen types of *armada* can be seen. Imagine the Saturday of a typical *batizado*, scheduled for 11.00am to 7pm, plus a party. At 10.30 there are five or six local students sweeping the floor of the hall, setting up a table to take money, laying out food for the *mestres*, the seats and instruments for the *bateria*, and greeting early arrivals, while a CD of *capoeira* music plays. *Bruxa* is there, helping set out the food. Around 10.45 other students begin to arrive, pay for the event or collect their Saturday wristband, change into *capoeira* clothing, deposit their stuff against the wall, and begin to stretch and warm up. *Bruxa* finds herself a chair in an unobtrusive space and starts to observe and record. She counts twenty-three students present. Now the first *armadas* of the day are done. There is plenty of space, and once people have stretched and warmed up, some begin to do a sequence of kicks. We stress that these are not directed by a teacher or any other person, they are individual. A Tolnbridge student, *Ikki*, arrives at 10.50. He kisses *Bruxa*, changes and leaves his belongings under her chair, stretches, and then moves into a space to undertake a sequence of all the *capoeira* kicks with each leg. He does all the circular kicks, including the *armada*, twelve times with each leg, as part of his preparation for the long (14 hour) day and night ahead. Around the hall others do the same. These are the first *armadas* of the day: done alone, and with plenty of space, to get the body moving.

At 11.05 the local teacher, *Sicinnius*, arrives with three visiting teachers who slept at his house on Friday night. *Bruxa* gets up and goes to hug all of them: students run to hug or shake hands with them, including *Ikki* who breaks off his routine to greet the teachers and then resumes it. The teachers change, eat some fruit, drink coffee or juice, then gather at the instruments to begin jamming on them or set out the clothing and CDs they have for sale on trestle tables. Two stretch and prepare to teach the first classes. The hall goes on filling up with students, including *Trovao* and other *Tonbridge* people. By 11.10 there are sixty-eight people. Those who have stretched either sit or stand talking to friends, find a space to do moves alone, or begin to play in pairs with other people.

These pairs find space and begin to play. If they are friends, flatmates, clubmates, lovers, these games are part of the enjoyment of the day, and part of warming up. In these 'pick up' pairs the second and third types of *armada* appear. In the games between friends and lovers the *armadas* are as friendly as the kick can be (it *is* an attack and the opponent must escape it). Men at the same belt level who know each other and each other's game may launch an attack.

Trovao and his flatmate *Bariaan* warm up in just such a way: their *armadas* are pleasurable exercise. They have trained together for five years, are on the same belt, and do not deliver their *armadas* in anger or malice, but as sensible preparation for the class that soon begins. *Bruxa* sees that game, and notes it.

Strangers playing each other are wary because they are unfamiliar with each other's belt level, playing style and conventions, and temperament: and because they have learnt that *capoeira* can be dangerous, injuries do occur, and, most importantly, no *capoeirista* can, or should be, trusted not to attack you. *Ikki* is now playing with a man from the same *capoeira* group but larger, and on a higher belt, from Poland. They too are warming up, and enjoying themselves, but *Ikki* is wary, because he does not know the stranger's game, nor his motivation. *Ikki* launches his *armadas* confident that the Pole will be able to evade them, expecting counter attacks because of the higher belt the man is wearing. His *armada* could be read as an aggressive move and provoke retaliation, or as 'normal', friendly, warming up, question and answer play. *Ikki* watches for the Pole's *armadas* to be launched at him: he has to evade them and decide whether or not to counter attack. He knows the Pole uses his *armada* to establish how good *Ikki* is and what sort of 'game' they are playing.

So now, in the hall, there are three types of *armada* going on. Some people are doing them alone to warm up. *Trovao* is doing them with a friend to get into the swing of *capoeira* play. *Ikki* is doing them warily to establish a relationship with a stranger. None is directed by an authority figure (teacher).

At 11.40 *Sicinnius* whistles loudly, then announces that the event starts in five minutes. People run to the lavatory, to take a mouthful of water, to stick plaster on blisters, strap support bandages onto knees, ankles or wrists, to peel off sweat shirts, tighten belts, and pull their long hair into ponytails or plaits. Anticipation rises. More students and teachers are arriving, as they continue to do all day. *Bruxa* gets up to hug people she knows, *Trovao* greets friends as they pass near him.

At 11.50 *Sicinnius* again whistles, seats everyone around the *bateria*, introduces the teachers present, (e.g. '*Mestre* Priam from Recife – claps for *Mestre* Priam everyone! – *Mestre* Priam came to our first *Batizado* in 2002, he is very welcome: a great teacher'). Once all the teachers are greeted, *Sicinnius* says '*Contra-Mestre Adonis* will lead the warm up', and then there will be three classes. *Adonis* will teach the children in the dance studio across the hall, *Contra-Mestre Endymion* will teach the beginners (no belts and first belts) at the west end of the hall, and *Mestre Admetos* will teach the advanced at the east end.

When the warm up is over and the three classes have separated it is 12.10. *Trovao* and *Ikki* go to take the advanced class, *Bruxa* moves so she can watch it. *Admetos* puts them in lines across the hall: there are forty-seven advanced students initially, twenty-nine men and eighteen women. By the end of the class at 1.15 it has grown to seventy-two, with thirty-eight men and thirty-four women, so the space available per student shrinks steadily during the hour. Initially there are eight students in the front row, and the lines are only three or four deep. At 1.00pm there are eleven in the front row, and the lines are six or seven deep.

During *Admetos*'s class four more *armadas* are seen by *Bruxa* and three of them done by *Trovao*. First, *Admetos* demonstrates and then drills the class in a short, simple, sequence of *ginga*, *armada* with the left leg, drop to *frontale*, up, *ginga*, *armada* with right leg, drop to *frontale*, up, *ginga*, which although done at speed, is easy for advanced students. *Admetos*'s demonstration of the kick is our fourth type: the pedagogic *armada*. It is normally delivered by teachers, but can be done by an advanced student asked by a teacher to demonstrate. In the pedagogic *armada*, which may be done slowly with accompanying commentary, the performer emphasises exactly how the kick should be done. So the demonstrator may say 'look where my arms are', or 'if you do *that* your kick misses his head' or 'put your leg *here* to have good style' or 'move your weight *here* to get speed in your kick' or 'protect yourself' or 'look in front' or '*don't* look at the floor' or 'spin your head *now*'. In an advanced class such pedagogic *armadas* are intended to improve the accuracy, or the style of the kick, or to help the student reduce their vulnerability to counter attacks by better players. Because it is an advanced class all learners (*discipulos*) can do the *armada*, but they all need to improve their style, accuracy, speed, and to build it into sequences of moves in a smooth, polished, sinuous way. That is the purpose of the master class. The pedagogic *armada* may be done sloppily, or dangerously, or in an ugly way, to demonstrate styles the teacher wants the student to *avoid*. Once *Admetos* has demonstrated the *armada* and its place in a simple sequence, and he only has to demonstrate it twice because it *is* simple, the 'drill' begins. The class does the sequence, doing their *armadas* whenever *Admetos* calls 'now' gesturing with his arm which leg is to do the kick. The class is drilled in this sequence 30 times.

Trovao and *Ikki* can do this sequence, these *armadas*, without physical or conceptual problems. It is well inside their skill range and easy to remember. In the second row, next to each other, all they have to be concerned about is not kicking those behind, in front or beside them, or being kicked by them. That responsibility grows as the numbers in the class rise and the space shrinks. This fifth *armada* is routine class drill, is exhilarating, raises a sweat, but only those hungover, ill, injured, or very out of condition, have any problem with this single, simple *armada* in communal individualised training.

Admetos then embellishes the sequence of kicks. He demonstrates and then drills *ginga*, two *armadas* with the left leg, cartwheel (*au*) to the left, two *armadas* with the right leg, *au* to the right. He drills that 30 times once again calling 'now' when the kicks are to be done. *Trovao* and *Ikki* now have to concentrate harder to avoid kicking or being kicked, and to ensure they move in unison. This is the sixth *armada*: a double in communal individualised training.

Admetos moves the class on to a third set of *armadas*. The player delivers an *armada* with the left leg and then immediately one with the right leg (i.e. spinning in the opposite direction). This is still physically and conceptually easy for *Trovao* and *Ikki*, but in the lines, in this class, it is increasingly hard not to be kicked, or to kick someone else. In this sequence, if a person's timing is wrong they are spinning the 'wrong' way. Again thirty of these (the seventh) are drilled.

There is a thirty second water break, then *Admetos* forms the class into a circle, and calls up *Professora Circe* who came with him from Olinda. *Admetos* says he wants the class to practice the kicks used in the *alto ligeiro*, the high fast game which is good for performances because it is showy.[7] He and *Circe* then demonstrate a sequence for the students to practise in pairs: A and B *ginga*, then A does a single *armada*, B escapes, does a single *armada*, A escapes, then does two *armadas* in the same direction while B escapes both, then B does two *armadas* in the same direction while A escapes them. Then both *au* to move round the circle of play, and they repeat the sequence kicking with the other leg (i.e. all the six kicks in the first sequence were with the left leg, now all six are with the right leg). *Admetos* and *Circe* demonstrate that three times, twice with *Admetos* starting the sequence, once with *Circe* going first. He tells them to 'get your friend and train. Be careful. It's crowded. Get space. Go slow. Speed up when you are both ready'. This is the eighth *armada*: still part of a training routine, but now in a question and answer sequence. No one *should* be fighting, or even trying to outplay their partner. The aim is to learn the *alto ligeiro* cooperatively.

Bruxa, who had joined the circle to see the sequence demonstrated, retreats rapidly to the edge of the room. There are now fifty-four students in the advanced class, and they pair up and try to find space to do the sequence. Several pairs go out of the fire doors into the corridor.

Trovao works with *Lunghri*, a clubmate at the same belt level. *Ikki* pairs up with *Crocus*, a girl from the Burminster club who has come with *Contra-Mestre Hercules*. *Ikki* has seen her at a festival, but never played *capoeira* with her. She is about his height, but he does not know how *Hercules*'s belts are awarded, and therefore has no idea how 'advanced' a *capoeirista* she is.[8] They shake hands and exchange names (real and *capoeira*). *Crocus* asks if *Ikki* knows what they are supposed to do, as she could not see the demonstration properly, or hear the translated instructions over the noise of the music, and the thumps and shrieks of the beginners. This is normal. Classes are noisy, and it is rarely easy to see demonstrations clearly. People check with their playing partner exactly what the sequence is before they begin to train in it. *Crocus*'s enquiry is being duplicated all over the hall. *Ikki* summarises and demonstrates slowly: 'my *armada*, your *esquiva*, your *armada*, my *esquiva*, my double *armada*, your *esquiva*, your double *armada*, my *esquiva*, *au*. Change leg'. *Crocus* repeats the sequence, and mimics the moves: says 'ok', and begins to *ginga*. *Ikki* picks up the *ginga*, and they begin to train. Both are aware of other pairs all around them, but they now focus on each other's kicks and escapes, matching their own moves to their partner's. Each of them is careful: the aim is to learn this *alto ligeiro* routine, to practise it, and to benefit from *Admetos*'s instruction.

However there are other things going on as they train this, eighth, *armada*. Each wants to impress the other with their skill and competence, their usefulness and niceness as a training class partner, and beyond that, uphold the 'honour' of their teachers *Achilles* and *Hercules*. An incompetent performance in the early stages of a class would not reflect well on the group each 'represents'. After about three minutes, *Admetos* calls out 'Mix'. *Ikki* and *Crocus* shake hands and find new partners. This time *Ikki* joins *Shere Khan*, a member of *Hercules*'s group who has been working in Tolnbridge for two years and training with *Achilles*. He is taller than *Ikki* and has a higher belt, but *Ikki* has trained with him and played him in *rodas*. He has only just joined *this* class, because he overslept. He and *Ikki* shake hands, and again *Ikki* is asked by his training partner to explain the sequence. *Ikki* does that and they start to play, slowly at first, then faster. Because *Shere Khan* is a tall man, *Ikki* does not have to get so low in his escapes, but Shere Khan's *armadas* are faster, and harder than *Crocus*'s, so *Ikki* needs to escape with greater urgency. *Ikki* does not have to 'represent' *Achilles*'s group to *Shere Khan*, but does want to practise the sequence with a person a bit better than he is.

Meanwhile *Trovao* is now training with *Crocus*, who plays him much as she played *Ikki*, because she can 'see' he is from the same group with the same level of belt. *Trovao* does not expect any violent aggression in *Crocus*'s *armadas*, because British girls do not usually practise *capoeira* as a fight, but he too wants to present himself as a competent represent-ative of *Achilles*'s group, and a pleasant training partner. Both *Trovao* and *Ikki* are aware of other pairs training very close to them, whose *armadas* could hit them by accident. After about two minutes *Admetos* calls them back to lines. *Trovao* and *Ikki* shake hands with their part-ners and take their places again. *Admetos* moves on from *armadas* to combinations of *armadas* and *queixadas*. Here the ninth *armada* can be seen: as a solitary training exercise, but in combination with a different kick.

Time passes. The class ends at 1.15. At 1.25 the advanced have a class with *Contra-Mestre Hercules* in which the *armadas* used are essentially 'repeats' of those taught in *Admetos*'s class. Then *Scinnius* announces that there are *rodas* so everyone can play. *Capoeira* games take place inside the *roda* (a ring of seated or standing people) who clap, sing, and focus their energy onto the two *capoeiristas* playing. Games last for a minute or so, and then the players retreat into the circle to be replaced by a new pair. At 2.10 there are eight small *rodas* all over the hall, with 10–12 students in each. *Bruxa* sits in one, clapping and singing, watch-ing *Shere Khan* play his friends from *Hercules*'s group. The teachers move from *roda* to *roda* playing each other, and students of various levels, advanced, beginner and children. In these *rodas* four further types of *armada* can be seen, in real games.

Adults in a *roda* with children do careful, slow, telegraphed *armadas* which passes safely high over their heads (even if they fail to do any kind of escape). Adult students playing other students practise their *armadas* in real games, representing their group, displaying their skills, preparing for their *batizado* or graduation later that day, trying to beat their friends and strangers, to enjoy it and not make fools of themselves. Students playing teachers use the *armada* when they are not escaping the kicks from the teachers, which is their major pre-occupation. No one wants to make an idiot of themselves playing a teacher, but if you have a good *armada* it is a kick you can use to put up a good show. *Capoeira* etiquette means that students do not attack or try to take down teachers in normal play, so only an ignorant or foolhardy person would try to hit a teacher with his *armada*.

Teachers playing teachers are professionals, whose *armadas* are not only better than any students', they are used with confidence that the opponents can and will escape. Their *armadas* (the thirteenth type) are higher, faster and more deviously delivered than those of students and may be used to compete with, or even fight the other teacher (although fights are usually stopped by the host or the senior *mestres*).

After a meal break the event moves from the big sports hall to a theatre, where the display (by teachers and advanced students for friends and family) and the *batizado* are to be held at 4.30. In the performance choreographed *armadas* (the fourteenth type) are exchanged between pairs of players, matched by the organiser for height, and all are highly skilled. These *armadas* are designed to make the audience gasp. This is movement as performance and performance as movement. At this event *Bruxa*, *Trovao* and *Ikki* are all spectators, because they are guests at *Scinnius*'s event. At equivalent performances staged by *Achilles*, *Trovao* and *Ikki* would be on the stage doing these *armadas*.[9]

At 5.00 the *Batizado* and Graduation begins. During the *capoeira* play that leads to the award of the *cordas* three further meanings of the *armada* are apparent. *Trovao*, *Bruxa* and *Ikki* would, again, be spectators at this festival: and indeed *Trovao* and *Ikki* might even leave to shower and change for the party because it is not their festival. If they stay to watch they observe the remaining type of *armada* from an informed position: they have played *mestres* for their successive belts and can empathise with the students being tested. In the ceremonial games that make up the *batizado* and the graduation, students play one or more teachers. Novices play one teacher, and try to display their best moves, including *armadas*, before the teacher ends the game, perhaps by taking them down. At a big festival there can be ten to thirty children getting belts, then twenty to fifty adult novices being 'baptised' so the process is a long one, with many *armadas* of varying competence. The *armadas* are the best the student can do, or is allowed to do by the teacher. This is the *armada* as a display of adequacy: the student shows she can deliver and escape one of the basic *capoeira* moves.

Students going for higher belts usually have to play several teachers. We have rarely seen more than fifteen people getting the second belt, and beyond that there may be only two or three 'candidates' for each promotion level so there is more time for the individual games between learners and experts. They use *armadas* as one of their moves, and will try to execute them stylishly, with grace and skill, in a smooth sequence of attacks and escapes. However as they play successive teachers, and begin to tire, the style and the smooth transitions become more ragged. These are *armadas* for display, delivered to signal competence in a basic move, but are embedded in a game, or series of games, and are recipient-designed. Students going for higher belts are able to deliver and escape *armadas*, just as they can play the *pandeiro* (tambourine).

These *discipulos* know that an audience of up to 100 people, many experienced *capoeiristas*, are watching them, scrutinising the quality of each kick, each escape, each acrobatic flourish, and the overall style of their game. These *armadas* are the best kicks delivered in a *roda* by the best students, and they represent the student, her teacher, and her group.

Finally, at a *batizado*, there are the *armadas* delivered by the teachers who are testing the students. They vary in speed and intensity according to the level of *discipulos* being baptised or graduated to a higher *corda*. The main meaning of the teacher's *armadas* is to display mastery of his own body, his mental and physical superiority over the student, and his care for their welfare. We have discussed *capoeira* teachers' bodies as objects of emulation elsewhere (Stephens & Delamont, 2006b). Teachers playing children for belts do slow, careful *armadas* high over their heads, which signal to the audience, especially the parents, that they are not putting the child in any danger. This is an *armada* to display the teacher as a caring, careful adult, who can be trusted to watch out for defenceless children. Teachers playing beginners use the *armada* to 'test' students' ability to escape simple spinning kicks, and, at the same time to show the audience that they can abort or halt *their* attacks if the novice does not dodge in time, or, tries to escape into the path of the kick rather than away from it. Novices often get confused, and escape the 'wrong' way. Playing novice adults, or those going for the lower grade belts, can provide teachers with the opportunity to make a fool of the student and get laughter from the audience. It would be possible for an expert to use the *armada* as part of a performative joke or a trick at the expense of the student. This deception or trickery (*malicia*) is fundamental to *capoeira* (Lewis, 1992) but is beyond the scope of this paper.

Teachers testing advanced students, especially those going for the highest 'student' *cordas* who have been praised by the host teacher for their commitment, loyalty and hard work, use the *armada* differently. At most *batizados* the host teacher has one or two loyal, long-serving students, well known to all the locals, going up to an advanced belt. Typically the announcer or the teacher says something on the following lines.

Finally we come to the pink and purple belt: tonight there are two students going for that belt. Rikki and Petunia. They were in my first class when I came to Forthamstead in 2001, they took their green belts in 2002, and they have trained hard ever since. They have worked really hard to learn *capoeira*, they are loyal, they are my friends, they have taught many classes for me, run the programme for the youth club, and planned all the parties for all the *batizados*. They are good *capoeiristas*, and today they go to the highest student belt in our group in Europe. They will play all the *mestres* in an *angola* game, and then in a *regional* game, and these will be *hard* games because they are the best students. I have asked Juniper to call an ambulance for them. Hey guys, claps for the *mestres* who will play Petunia and Rikki.

In the games following such an announcement, the teachers play the students almost as if they were fellow teachers: the *armadas* are 'real', and are the advanced students' *reward* for their long hours of practice. They, and the audience, know that the teachers are using the armada as part of real attacks that they must escape and counterattack, because that is what they have trained for. As an advanced student said to *Bruxa* at one of *Perseus*'s festivals 'That's why you train, so you can face them'.

Discussion

We have set out eighteen possible 'meanings' that can be attached to one kick, all of which are routinely on display during any *capoeira* festival. Any other *capoeira* move, whether attack or escape, could have been used in the same way, because all movements have multiple meanings. Observing these multiple meanings, and in *Trovao*'s case, performing thirteen of them, in private practice, in lessons, and in *roda* play, involves both of us in mobile research, on a set of mobilities. Our research aim is to understand *capoeira* as it is taught and learnt outside Brazil. To attain our goal we have, out of necessity to move ourselves. We both have to be able to see the teachers' pedagogic *armadas*, to appreciate the displays, and to avoid being kicked ourselves.

Clearly research on *capoeira batizados* and on *capoeira* itself is very specific. However the research is an example of key features of Urry's (2007) mobilities paradigm in action, and can therefore be used to explore general features of that paradigm. Urry (2007: 40–43) sets out nine 'mobile methods': that is 'methods on the move' needed to implement the mobilities paradigm. These are the study of transfer points or liminal places; of places that themselves move (such as ships); of objects that can be followed around; of objects that can be followed around; of memories of movements past; of imagined and anticipated movement; of virtual movement (through for example blogs); of time-space diaries; of moving informants by moving with them; and finally of observing moving bodies. Urry's book illustrates these nine methodological imperatives with a variety of projects. Our contention is that our ethnography could equally well illustrate Urry's schema.

Focusing upon the *batizado* is certainly a study of transfer points, conducted by two ethnographers moving with their informants and studying moving bodies. The potential of those three methods for achieving Urry's paradigm has been the keystone of this paper. We have placed less emphasis on Urry's other strategies, but they are implicit in the data on *batizados*. Students and teachers have memories of *batizados* that form the topic of many conversations for years afterwards and are preserved on DVDs made at the events. Before *batizados* they are anticipated and imagined (indeed such anticipation motivates routine training). Experiences at *batizados* are the subjects of numerous blogs, emails, Face Book pages etc., and of time-space diaries. The last two of Urry's nine approaches may seem absent from the *capoeira* ethnography presented in the paper. In fact they are two directions that our research should take. In future it would be useful to explore *batizados* as moving events, and tracking moving objects in *capoeira*. Among the moving objects are the instruments used in the *bateria*, and focusing upon these would be an illuminating way to study *capoeira*. Additionally as each *capoeira* group has clubs and classes across the world it could be argued that the *batizados* of any particular group – be it *Cordao de Ouro*, or *Beribazu* are moving, like ships. During a year it would be productive to 'follow' the dynamics of a *capoeira* group by researching the *batizados* in different countries over a temporal period. *Beribazu* for example has *batizados* in Britain, Poland (Reis, 2003, 2005) the USA and several different Brazilian cities. An understanding of *Beribazu* would be enhanced by the study of such a moving phenomenon. We see our ethnography therefore as an example of Urry's ideas in practice, and as a route map for further research on a mobile phenomenon.

Acknowledgements

We are grateful to Mrs Rosemary Bartle Jones for word processing the paper, to all the *capoeira* teachers we have learnt from, especially *Achilles*, to the many students who have shared their *capoeira* experiences with us, and to colleagues in Cardiff who have commented on our papers. All the names, of places and people, are pseudonyms.

Teachers have pseudonyms from Greek and Roman mythology and history (Harmodias, Persephone), male students' pseudonymous nicknames are from *The Jungle Book* (Raksha), female students' pseudonymous nicknames are flowers and trees.

Notes

1   *Trovao* does *capoeira*, *Bruxa* watches it done. We then discuss what we are learning as an embodied practice (*Trovao*) and an observed one (*Bruxa*). As members of a successful *capoeira* group we have real *capoeira* nicknames not used in our publications to protect our informants.

2   Data on twenty *Batizados* were collected as follows. Nine were attended by both *Trovao* and *Bruxa*; eight festivals in Tolnbridge and Cloisterham run by *Achilles*, *Trovao*'s teacher, and the instructor we have studied most intensively. Data have been gathered at *Perseus*'s Longhampston club by both authors once (in 2004) and *Bruxa* on four other occasions. Additionally *Bruxa* has observed *Batizados* run by *Leontis*, *Hermes* and *Ajax* in London and *Hercules* in Burminster. *Trovao* attended *Achilles*'s first ever festival in the UK, when he got his own blue (first) *corda*, *Bruxa* did not.

3   There are two types of *capoeira*: *regional* which is played faster and more upright with kicks roughly similar to karate, and *angola* which is played more slowly and closer to the ground. Many UK *capoeira* teachers offer both styles, but purist *angola* clubs do not have a system of belts (Assuncao, 2005).

4   There are many different groups in *capoeira*. A student is baptised into her teacher's group. The London based Brazilian, *Mestre* Poncianinho (who appeared in a station ident on BBC One, in a Harry Potter film, and in several TV commercials) belongs to *Cordao de Ouro* (Golden Cord) and anyone baptised by Poncianinho is therefore a member of *Cordao de Ouro*. Different groups have different belt systems: a blue belt in one group is a beginner, in another someone with ten-years' experience.

5   There are women teaching *capoeira*, including distinguished *mestras*, but our data are about male teachers.

6   *Professor* is the teaching grade below *Contra-mestre*, above *Contra-Mestre* is the *mestre*.

7   The real *alto ligeiro* is done without escapes, but a teacher faced with a crowded class full of students he does not know would train the kicks *with* escapes for safety's sake.

8   Advanced classes in the UK contain people who have their second belt (three years' experience) right up to learners with ten years of hard training.

9   At *Achilles*'s 2009 summer festival *Bruxa* acted as continuity announcer, 'explaining' the performance to the audience.

7
Being There/Seeing There: Recording and Analysing Life in the Car

Eric Laurier

There will be two mobilities at work in this chapter, the first being the mobility of the research subjects and the second being the mobility of the video data they produced. There will only be the space to touch on the topic of mobility given our orientation is to methods for its analysis. However, studying mobile cultures raises wider questions about how to engage with and respond to video as a form of data in the social sciences. These are questions about observation, authorship and confidence.

The project which utilised the methodology described in this chapter was part of a thirty month study of social interaction inside cars.[1] From fifteen vehicles, approximately two hundred and forty hours of footage was gathered of commuters, car sharers, families and friends travelling. What I would like to do rather than moving on to provide a digest of the current literature is to take us immediately to the mobile research site and a group of research subjects setting up the camcorders inside their car.

14 Famous Moonie

A group of hill-runners who car share. They are about to return home after a hill-run. Driver is pushing camcorders into place on dashboard. E is the driver, FP front passenger and RP is the rear passenger.

FP: ...Simon!
       ((others laughing))
FP: That's it, I'm in now
RP: I hope they're not filming you
       ((laughter))
E: Naked men in the back
FP: ((turning round to rear)) Can you imagine though if you forgot

it was on and you, you were on the job and then ((returns to looking toward front)) Can you imagine them at university ((returns to rear look)) 'Bout doin it. Come on ((slaps thigh)) let's do it Elsa for a laugh ((looks across at Elsa)).

((returns to looking at rear)) Can you imagine them all sitting there taking notes and discussing it. 'What they, what they doin now?' ((pulls on seatbelt))
((laughing FP and RP))

E: ((Scratching midgebites on side of head)) Yeah you're supposed to forget the cameras are there, just
FP: ((laughing louder)) What if ((stops talking and grabs nose))
E: Yeah it's not often we have guys changing in the backseat
FP: Oh Christ! Is it on just now?
E: No ((reassuringly))
FP: Why's the red light on?
       ((FP laughs, E looks across, FP looks around))
RP1:   Oh Christ! ((FP looks back at him))
       ((covers rear view of RP's groin))

E: I've got to get a girl in there in the back as well. Sarah might
      want a lift. Well, Sarah want a lift to the pub ((rear passenger
      turning around to moonie camera))

FP: And [where's Sean] going?'

RP: [David]

E: Oh Sean's [going] for another run

RP: [David]

FP: Has he ((catches sight of moonie)) Ha! ((recoils)) Paul!

RP: ((moves back to seat))

FP: Jesus Christ Man!

      ((laughing))

FP: Christ all bloody mighty!

RP: Apologies Hayden,[2] couldn't resist ((laughter))

FP: Prob, probably won't be Hayden that sees it.
      ((laughter))

What indeed did we, earnest scholars, 'taking notes and discussing it' make of all this. As the passenger imagines us, academics, asking 'What are they doing now?' In responding to the original video footage and to this transcript I find the tidy dichotomies of methodological planning and moral reflection knocked sideways. Also, as is so often the case, I find myself enjoying the records left of the lives of others, others who are as concerned with their enjoyment and the future enjoyment of the record. With a smile settling on our face after being moonied by those we would seek to study and a distinctly social scientific blend of humbleness and arrogance in the face of their resistance to easy explanations we can now begin to ask ourselves 'what are they doing now?'

Having located the methodology as much in the hands of the unruly and ruly occupants of a car, as in the research team usually assumed to implementing it, we can now move to more familiar ground and consider some of the displacements and re-arrangements of ethnographic knowledge that come with the use of video. Clifford Geertz has argued quite persuasively that we are willing to take seriously the ethnographer's knowledge of other cultures because they have, for a while, become part of another form of life, that they have, in short, "been there" (Geertz, 1988: 4). We accept what the ethnographer has to say about witchcraft amongst the Azande, reliability in the NHS or inspection routines in UK residential high rises because they are an informed and transformed observer on what has happened and continues to happen in these elsewheres. Ethnographers gather, garner and frequently stumble over knowledge of other cultures through their *presence*. My question is: what happens when the ethnographer brings back video records from *there* to *here* and in doing so shifts some of the burden and responsibility of witnessing to a new audience over here?

Just over a decade ago ethnographers first began to tease out the peculiarities of studying mobile cultures and cultures of mobility (Clifford, 1997; Marcus, 1995). These are cases where our presence 'there' is located as a troublesome matter from the outset. One common solution to learning about various mobile practices has been a number of different modes of travelling *with*, be they walking with munro-baggers or reindeer herders (Lorimer, 2002, 2005, 2007) or riding along with snow ploughs (Weilenmann, 2003) or sitting beside bus drivers (Normark, 2006) or cycling with bikers (Spinney, 2006) or sitting with other rail passengers (Watts, forthcoming). In the Habitable Cars project our similarly simple solution was to cadge a lift in our study-vehicles on a number of their typical journeys. For each of our study vehicles myself, or one of the other project members, travelled with them for a week or so, becoming passenger-seat ethnographers. A favourite saying amongst ethnographers (Fielding, 1994; Rouncefield, 2002) is the Native American credo that 'one should never criticise a man until you have walked a mile in his moccasins'. For the cars project this became 'one should never criticise a car owner until you have sat for a week in her passenger seat'.

We had a notional two-part 'follow and film fortnight', that usually ended up stretching over a month. The second part, that followed on from the passenger-seat ethnography week, was the ethnographer handing a camcorder kit containing two camcorders with their related lenses, cables, batteries and tapes and two foam cubes (visible in the first still of the transcript) over to the project participants. After talking over with the travellers how each camcorder worked, how to fit the fish-eye lenses, change and charge the batteries, we would look around the interior working out where the camcorders could be positioned in the car. Finally, when the car was ready to go, we would say things like:

Hayden naturally

Hayden (project member and hill-runner) has installed the pair of camcorders into Elsa's, the driver, car for a first attempt at filming their journey. RP1 & 2 are the rear passengers and FP, the front.

H: On

E: Smile now ((looks back to rear passengers))
RP2:   Little red light's [on]
            +
E: [yeah]
RP2:   So I expect [so]
            +
H: [can you see a red light]
Several: Ye[ah]
            +
H: [yup] Okay well let's hope ((withdrawing from camcorder))
            that's on and this'll be a little test run on the way out
RP1:   Okay
H: Uhm, [fingers crossed]
            +
E: [okay] See you [there then]
            +
RP2:   [okay]
H: Naturally-occurring social activity that's what we're looking for
E: Naturally-occurring
H: ((laughs heartily))
RP2:   (he's) been doing [that while]
            +
RP1:   [There's] been snogging in the back
H: ((laughing heartily at that))
RP2:   Oh yeah
H: So what way you going then Elsa, out the back? ((points))
E: That way. The cut-off. Yeah
E: ((looking to camcorder)) Watch that doesn't fall.
            ((chat in rear))
RP2:   Derek your job is to hold that thing in place
FP: (2.0) ((looking at camcorder then to rear)) And what about
            the camera
            ((wheeze-laugh))
RP2:   Right
            ((driver turning car around, passengers following her
            manoeuvres))

Along with Hayden's mildly ironic reminder to the occupants of the
car that the project wants 'naturally occurring social activity' the most
heavily used instruction from us, because of its less obscure nature,
was that we wanted 'typical journeys'. We emphasised to our driver
and passengers that there was no need to entertain us in the style of

docu-soaps or reality television shows. Alongside our verbal requests for mundane, humdrum journeys, we gave the participants FAQ sheets, going over once again our desire to have them do what they ordinarily do.

From the set-up of the camcorders and from seeing the first and second clips it should be apparent that how we went about recording life in the cars was anything but covert. It was, as Lomax and Casey (1998) para 4.1, put it: 'a product of the occasioned activities of the researchers and participants'. Rather than seeing the presence of the camcorders as contaminating the video record of mobile practices we exploited them, as Lomax and Casey went on to argue, as 'a resource for exploring the interactional production of those activities'. That said, the unattended camera's presence is quite a bit less distracting and disruptive than having a chin-rubbing character with a notebook busily scribbling in the passenger seat. The 'there-ness' of the ethnographer is re-distributed because it has been, in a Latourian (Latour, 1992) sense, delegated to a mute machine, a dumb digital device. Using the camcorder we can be present to catch whatever is happening while not making an overly disruptive difference in what is happening.

Even with the awareness that the video record is a product of the ethnographer and their participants, and, even when you are not seeking to justify its claims to knowledge, at the point that the research team, or any other researchers, sat down to watch the recordings there was, and is, still a tremendous temptation to slip into an *observational* mode. When we scrutinise video footage of mobile practices and when we talk about its strengths as empirical material, at that time we risk becoming the disengaged and detached neutral observer that positivist inquiry requires. In watching the video-recording, we do not prompt the practices, it seems that there is only a steady flow of information *from* the footage *to* the researcher. No information is passed back from ethnographer to participant once this form of self-filming is underway. How could it be? The events are in the past. The ethnographers are not even in the car and thus have become an observer so withdrawn that they become an 'absent presence' (Raffel, 1979). Once the events that occur during any car journey have been taped, in reviewing the tapes the project members cannot intervene or in other ways change what happens in them. The video footage seems to return the participants' local organisational practices to their source.

In this observational mode we contemplate the re-use of video footage, the sharing of stable standardised data, without the need or requirement that we understand what its maker was up to. In sorting through the video footage we could isolate events made by the maker from those made by the culture being recorded. Exercising the latter sort of criterion I have selected out these first two clips. The remainder once such a removal has been carried out should leave clips that could be analysed without reference to this specific project. By enrolling the car travellers into the recording process we seem to transfer accountability for the recording. 'The responsible observer is one who can make what he observes responsible for what he records' (Raffel, 1979: 29). It would appear that thereby we escape Geertz's attribution of responsibility to the author because the subjects are becoming the authors of their own footage.

Such are the temptations of the relationship between an observational mode and the video record. What of the event and the video record? Here the camcorder appears to offer the possibility of an absent-presence at an event, while equally events are apparently disclosing themselves to the camcorder without any need for the ethnographer to do anything. What Geertz's (1988) work was constantly reminding us of, was that ethnographers were authors, that each one had different styles for securing their authority. The camcorder, then, seems to promise that idiosyncrasies of note-taking, documentation and diary-keeping might disappear, to be replaced by the impassive standardised recording of the digital camcorder. Events disclose themselves to the camcorder, they are not summoned or directed along the way by the ethnographer's presence.

Let's change tack. Reading Harvey Sacks's lectures (1992) not long after I completed my PhD I felt a sense of revelation at a quite distinct sense of intellectual ambition and analytic craft.[3] Here it seemed were discoveries and solutions in the face of seemingly intractable problems from social and cultural theory. Sacks's lectures were rich in findings. He would analyse a single transcript and show how particularities began to unpick generalities developed through the emaciated examples of social ordering worked on by theorists such as Talcott Parsons. Examining, famously, turn-taking mechanisms (Sacks *et al.*, 1974), along with categorisation practices, pauses, grammar, concepts of mind, speaker-selection techniques, enforcements of order, the accomplishment of ordinariness (Sacks, 1984), laughter (Sacks, 1978) and the list goes on and on, Sacks lectures are an embarrassment of riches. After reading them you notice all manner of missed ordinary accomplishments afresh.

Excited by the promise of audio-recordings and their pain-staking transcription, Sacks's desire was to found a truly scientific science of human behaviour in the form of conversation analysis (Lynch, 1993). Mike Lynch reminds us that, as his work developed and departed from its ethnomethodological origins, Sacks profitably took up a mistaken version of what scientific method is, in order to place his studies on a firmer foundation. Garfinkel's (1967) ethnomethodology is infamous for its constant excavation and explosive demolition of social science foundations of various sorts. Sacks accepted that a key element in the establishment of scientific findings was a technology of observation, data, replication of findings and professionalised analysis. The ubiquity of the phenomena (speaking) and the tape recorder and the transcript would allow the creation of stable data, checking and replication of results.

To echo Lynch's remarks, Harvey Sacks's studies of language had a preceding, and then accompanying, 'natural philosophical' shape (Lynch, 1993: 216). In his early lectures he challenged the existing social science studies of language in a manner similar to that of J.L. Austin's criticisms of philosophy (Cavell, 2002). For Sacks the tape-record was the place to begin and to, in an ethnomethodological vein, re-specify various classical topics in social science methods such as observation, description, models and explanations. For those who have taken up Sack's legacy, like myself, recordings of various things happening, are at hand to be returned to again and again. The care exhibited in this return is one that we usually find in literary scholars who return again and again to canonical works such as the plays of Shakespeare or, equally, the film Alien (Kuhn, 1990; Mulhall, 2002).

To bring this back to the day-to-day work of the Habitable Cars Project, the twenty vehicles recruited for the project generated over one hundred hours of video which we archived into six hundred and sixty indexed clips. In editing out the six hundred and sixty clips we were pursuing once again 'typicality', identifying general features, conventions and, equally, exceptional events. Much like the manner in which the process of transcription has long been recognised as the beginnings of theorising conversation, so it is that the editing out of events from the long run of a journey is part and parcel of the project's inquiries. In effect, more than three quarters of the footage disappears at this stage, edited away to the virtual cutting room floor. That the larger part of the original data, be it questionnaire sheets or audio-recordings of interviews, disappears is surely not a surprise to those pursuing substantial research projects. What can compound this disappearance in quantity is the loss of time as researchers become engulfed by the requirements of indexing and coding their data-sets. The danger of trying to put on show every event and action that the project recorded is that we become bureaucrats of data-sets rather than analysts of social practices. As a consequence of this desire we have a changed set of evidential objects and criteria to the example and the illustration; we have instead the fragment, the specimen, the actual and the typical.

You have seen two fragments already neither of which was analysed at any great length because they were there to form a dialogue from the participants with our concerns as social scientists. What we will move on to now, briefly, is how the analysis proceeds of the following fragment:

$P = passenger, D = driver$[4]
      ((*Approaching slip road*))
      ((*P raises his hand to point at car pulling out*))
P: (inaudible)
      ((*His finger then touches nose*))
P: I thought he was gonna..., ((*looks into car as they pass it*)) aye, she was gonnae pull out and go for it
P: Somebody pulled out in front of me
      +
      ((*turns toward driver*))
      (1.0)
D: ((*turns toward passenger*))
P: Comin' in
      ((*both return to looking forwards*))
      (1.0)
P: Just down from your street ((*sideways nod*))
      (0.5)

P: ((*shakes his head*)) Didnae even apol'. I beeped at him and every-
       thing he didn't bother apologising ((*looks toward driver extendedly
       and with additional move into driver's space*))

P: ((*looks away out of passenger window shaking his head and then
       ahead*))
       (3.0)

D: ((*moves head slightly toward passenger*)) It'll be your fault of course

P: ((*looks across*)) Mhm?

D: ((*looks and meets P's look*)) It'll be your fault
       ((*both turn away*))

P: It'll be ma fault yeah ((*both smiling*)) Oh yeah ((*looks out passenger
       window*))

This was one of the first clips we worked with and in a sense we could
have worked on it for the whole 30 months of the project. Conversation analysts have worked on single transcripts for that length of
time. What happens in the beginning of the clip caught my attention,
as it did the passenger's, for what appears to be a getting-too-close
shave. An accident almost happens in the way that accidents quite
often almost happen on the roads. At the time I was wanting to respond to the fact that a great deal of driving is studied through accident
statistics. Here was a moment that is invisible to those statistics, perhaps initially invisible to the driver, noticed by the passenger and
brought to the driver's attention.

Typical of the interest of ethnomethodology we find convergence
here between our work as viewers of the video looking for events happening and the passengers themselves noticing something happening.
The passenger is in a risky position in terms of pointing out something
that is happening since as a passenger one should not be interfering in
the driving. However that was not quite the way that our analysis
developed as we re-watched and discussed what was happening in it.
We become more and more interested in the accountability of the
gesture, that its consequence is dealt with by saying 'I thought he was
gonnae pull out'. In trying to engage with cognitive studies of driving
as an ideal case between thinking and habit, our interest was piqued by
the ordinary usage in this clip of 'I thought'. So we find ourselves sur-
prised by where we ended up and how passengers and drivers maintain
reason and order on the road and inside the car (Laurier, forthcoming).

Proponents of video (and this is not just CA) are excited by the
opportunity to show and share clips such as the one transcribed above.
The consultation of the original recording appears to provide for other

researchers to re-check the analysis or to argue with the descriptions made by any particular analyst of the events recorded on video. Here is where we return to the observational mode discussed earlier: the presence of the ethnographer as an opinionated observer is lessened by their substitution by other analysts when the video record is shown during a data session. Cross-checking with other researchers will secure the validity of the knowledge gathered from the video material. What we might forget in saying so is that although literary or film studies share a common text they are happy to disagree about any novel, play or feature as much as they agree about it. Their readings of a poem or a novel are embedded in the opinions they wish to venture (Livingston, 1995). Yet what they perhaps display in all of their disagreements is a willingness to return to a work that they are reading together, and to keep returning to that work. How often in the social sciences do we find a community of researchers willing to keep returning to an interview transcript or video clip? Perhaps not so much to get to the bottom of it, to say for once and for all what is really means but to? To what? What is the point of looking at a video clip or transcript of some quite ordinary thing?

Clifford Geertz (1988) described Malinowki's style as 'I-witnessing'. Compared to the slightly uneasy travelogue of Levi Straus, Malinowski's is a total immersion in the 'there'. By Malinsowski's engulfment we also come to realise there is a lot more to the 'there' than one would expect from other forms of ethnography where the identity of the ethnographer is never in danger of being drowned out by their immersion in another culture. Geertz argued that what we have in Malinowski's work is the comprehension of the self through the detour of the other. To secure its claims to knowledge, the 'I' of the ethnographer carries ever more weight, since the experience of the encounter offers up self-knowledge as much as knowledge of another culture, and so the 'I' must become ever more convincing. As Geertz puts it, 'Erasing distance between observer and observed ... soaking it up and writing it down' (p. 83). In the project our aim was not however to push the car travellers into serving as an exotic other. They are, quite ordinary, familiar others.

What then is happening in the car project when its three researchers spend a great deal of time viewing video clips of typical practice? The site of immersion in the practices of others is displaced and the desire to be the stranger to those others or to become a stranger to one's self disappears. As Cavell puts it briefly, the video clips become a way of responding to the 'appeal of the ordinary' (Cavell, 1990: 66). Or to adapt his words to a more ethnomethodological spirit, between for instance, social science's sceptical attitude toward human action and the grammars of ordinary practice. In attending to video recordings, the distance of a sceptical social *science* found in the gap between observer and observed is posed in a different configuration and our intimacy in sharing in human practice and the implications of doing so are transfigured.

How I describe my video viewing methods to students are in terms, borrowed again from Harvey Sacks, of a style of 'unmotivated listening and looking'. I say so, in order to try and shake them free from the routine habit of applying one or another theory from the social sciences to what they are watching and listening to. Unmotivated viewing is a term not so unlike those in the social sciences. Moreover, it was developed from two eminent predecessors: ethnomethodological indifference and phenomenological intactness. Its trouble is that it does still sound too much like the uncaring, uninvolved observer and uncommitted observer of the other that a certain form of positivist observation wants. And in writing this paper my realisation is that I will have to stop offering 'unmotivated watching' as a method for analysis. However, what we are asking of ourselves is not, in the end, total immersion either. It is not as we might lose ourselves in what we are watching totally. Rather it is the *patience* of conversation analysis with ordinary practice that I find myself valuing.

Patience is necessary because footage, at the outset, is of an unremarkable and barely noticeable part of everyday life. Often the Habitable Cars project investigators would sit for an hour with a video clip just getting nowhere, stumbling around amongst dead ends, saying something we have said too many times before or wanting to jump to a line from a theorist we have been reading. When, and if, we were patient with a clip, something would emerge. What I have learnt from this is that we have to try and be confident with the mundane aspects of mobility. Confident, because what seems to be in danger is that we lose our confidence in ordinary practice to say anything back to grander theories of human affairs. We have to begin 'in the street, or in doorways' (Cavell, 1994: 57) or, as here, in the car where analysis of these ordinary practices will waken us to wonder. Using video will not allow us to ground our theories or anyone else's, despite Glaser and Strauss's famous promise that we could ground theory through empirical work. The promise of this sort of empirically guided study is that it may help give us a brief rest from theorising mobility.

Notes

1 For further details of this ESRC visit the project website, search via 'Habitable Cars' (URLs are likely to be updated regularly).
2 Hayden Lorimer is one of the co-investigators on the project.
3 Reading Cavell's reflections on his encounter with J. L. Austin has greatly helped me in understanding the experience of encountering Sacks if one has had a background in social and cultural theory (Cavell, 1994, 2002).
4 Denoting these person as these particular categories of actor (e.g. 'passenger') already begins to make assumptions about the relevant categories, where the fact that they are a passenger (or a man or a commuter) may or may not be relevant. I can only signal such concerns here, for a fuller examination of omni-presence or not of categories (H. Sacks, 1992; Schegloff, 2007).

8
Writing Mobility: Australia's Working Holiday Programme

Nick Clarke

I remember a young woman with an Irish accent and a Thai sarong draped over her shoulders. She was talking to a man with a Canadian flag on his backpack. They were competing with travel stories, and sharing pub-quiz knowledge of Australia and other places. In the background, a couple of young men wearing Premiership football shirts sang the Dutch national anthem. And a Japanese woman passed by dragging a bright pink suitcase on wheels behind her. I remember conversations about the future. Someone listed the places they wanted to visit before they died. Someone else detailed their own plans, which involved getting married, buying a house, having kids, and growing old in the same place they grew up in, surrounded by family and old friends. I remember a young man with a neuropathic bladder. He was reluctant to leave the country in which his condition was diagnosed. He worried that he might not get the catheters he needed as easily elsewhere. I remember news of a detention centre in the middle of the desert, where refugees from other countries were held behind razor wire for years on end.

Following mobility

In this chapter, I return to the period 2000 to 2003 and a research project on British working holiday makers in Australia, to offer some thoughts on the problem of writing about mobility – since methodology neither begins nor ends with data construction. This research project was inspired by a number of things, and by two texts in particular. The first was James Clifford's essay on 'Travelling Cultures'

(Clifford, 1992). In this essay, Clifford criticises the localising strategy of traditional anthropology, which locates culture in a particular field (the village, for example), studies localised dwelling and rooted experience, and marginalises travel, transport technologies, prior and ongoing contact and commerce with other places, and national and international context. He calls on anthropologists and related researchers to rethink cultures as sites of both dwelling and travelling, and to focus more sharply on travelling subjects (migrants, tourists etc.), places (airports, hotels, motels etc.), experiences (superficiality, transformation etc.), and products (stories, knowledges, traditions etc.). The second text was John Urry's *Sociology Beyond Societies* (Urry, 2000). In this manifesto for sociology and the social sciences more broadly, Urry observes that, in contemporary times, social relations are being reconstituted by inhuman objects (technologies, texts, images etc.) that produce mobilities across borders (migrants, risks, money, consumer goods and services etc.) with consequences for dwelling, citizenship and nation-states. Communities of propinquity, localness and communion are being replaced or at least supplemented by bunds, neonationalisms and diaspora. Citizenships of stasis are being replaced or at least supplemented by multi-tiered, postnational and deterritorialised citizenships. The role of states is shifting from that traditionally associated with gardeners (who nurture populations) to that traditionally associated with gamekeepers (who regulate and facilitate mobilities). These observations lead Urry to replace the undermined metaphors of region, society and sovereign nation-state with new metaphors of networks (sets of interconnecting nodes), fluids (people, information, money etc. which flow through networks), and scapes (the routeways of machines and organisations along which flows get relayed). He also calls for sociology and related disciplines to refocus away from nation-state-society and towards mobilities: the travels of people, ideas and objects across borders, and their effects on experiences of time, space, dwelling and citizenship.

Following these calls and others, I headed from England to Australia in November 2001 to follow the set of travelling people, ideas, objects and experiences connected through Australia's Working Holiday Programme. This programme dates back to 1975 when Australia introduced the universal visa system. Prior to this, all Commonwealth and Irish citizens of European descent were exempt from the visa requirement. As part of the new system, in order to preserve the existing arrangement under which Australians were permitted to holiday and work in the United Kingdom, the Working Holiday Programme was established, allowing young British citizens similar rights in Australia. The objectives of the programme are stated on Visa Application Form 1150: 'The Working Holiday Programme aims to promote international understanding. It provides opportunities for resourceful, self-reliant and adaptable young people to holiday in Australia and to supplement their funds through incidental employment'. In 2001, the regulations contained the following requirements. Applicants must be citizens of the UK, the Republic of Ireland, Canada, Japan, the Republic of Korea, Malta, the Netherlands, Germany, Sweden, Denmark, Norway or Hong Kong. They must be aged between 18 and 30. They must have no dependent children. They must not work for the same employer or study for more than three months. And they must be of good health (with no disease or disability likely to endanger or be of cost to the Australian community), wealth (with approximately AU$5000 for personal support during their stay and their return airfare home), and character (with no convictions and no mental illness). I return to these objectives and regulations below. The other thing to note about the programme is that, during the fieldwork year (2000–2001), over 75,000 working holiday visas were issued by the Commonwealth Government of Australia, almost 40,000 of these to British citizens.

This last point was one of the things that interested me about the programme. How had it come to this, that on any one day almost 50,000 young people from twelve different countries could be found not permanently migrating to Australia, nor taking a couple of weeks' vacation in Australia, but working and holidaying in Australia for anything up to 12 months (with the opportunity of extending this period beyond 12 months and up to four years with sponsorship from an employer)? Other things that interested me included the travelling and dwelling practices of working holiday makers, the relationship between these practices and discourses of international understanding and personal development, the relationship between working holiday maker practices and the travelling and dwelling practices of Australian citizens, and the regulation of working holiday maker practices by local, provincial and national governments of Australia. The findings of this research project have been published elsewhere (Clarke, 2004a, 2004b, 2005). Very briefly, I found that British working holiday makers in Australia practise life on the move by travelling-in-dwelling – periodically returning 'home' through websites such as Guardian Online, television programmes such as The Bill, radio channels such as the BCC World Service, music and photographs carried on their person, and through telephone calls, e-mail conversations, and gift exchanges with family and friends. They also practise life on the move by dwelling-in-travelling – carving out their own place in each space they stop at temporarily, by constructing traveller communities using mobile technologies of communication (web-based e-mail accounts, mobile telephones, listings magazines) and more traditional technologies of sociability (card and board games, alcohol, hostels with dorm-rooms and communal areas). These travel practices can lead to international understanding and personal development, if tightly defined. Personal development, when understood to involve both skills acquisition and self-narrative refinement (Giddens, 1991), is encouraged by the upheavals, liminal spaces, slow times, and inscription practices generated by working holidays. International understanding, understood as a cultural construct, is crafted by active and reflexive working holiday makers out of the heterogeneous spaces and ethnographic times of working holidays. I discuss other findings of the research in the sections that follow but the focus of the rest of this chapter is on methodology and, in particular, the writing part of research methodology.

Provocative mobility

The first thing to say about methodology is that, during this research project, I never really gave too much thought to the relationship between mobility and methodology. Certainly, I was not provoked by colleagues or referees, or by anything I was reading at the time, to consider such a thing as 'mobile methodologies'. The research proceeded from a philosophical position influenced by reflexive realism and critical constructivism (Delanty, 1997), post-Marxism (see Corbridge, 1989), structuration theory (Giddens, 1984), Actor Network Theory (see Law and Hassard, 1999), and pragmatic universalism (Albrow, 1996). Following two exemplary studies of mobile subjects and spaces published prior to and during the project (Ong, 1999; Smith, 2001), the research combined two modes of research practice: political economy; and an ethnography informed by Clifford Geertz's thick description (Geertz, 1973) and Michael Burawoy's extended case method (Burawoy, 1998). In turn, these two modes of research practice translated into four methods of data construction: desk-based contextual work; corporate interviews with representatives of Australia's backpacker industry and Australian national and local government; participant observation; and in-depth interviews with working holiday makers.

I was, however, provoked from another direction to give at least some thought to the relationship between mobility and methodology – from the direction of mobility itself. What I mean by this is that following travelling people, ideas, objects and experiences led me to certain epistemological positions and practical decisions that are different from those to be found with more sedentary research subjects. At the epistemological level, since mobility involves seemingly forever changing contexts and perspectives, and seemingly never-ending encounters with new fields, the tone of the project became more cautious and modest – an acknowledgement that no obvious end-point exists for research on circulation, and claims made on the basis of such research are freeze-frames from a world already on the move again. At the practical level, I would have liked to run focus groups with working holiday makers, convinced by Cook and Crang (1995), among others, that apparently individual thoughts and feelings tend to get worked-up through interaction within groups. But organising focus groups, and running them more than once with the same group so as to establish trust and rapport, is almost impossible when dealing with travelling subjects and their diverse itineraries. Indeed, completing the in-depth interviews was difficult enough, given the characteristics of travelling spaces which tend not to be quiet and clear of distractions.

Most importantly, for the purposes of this chapter, the mobilities I encountered in the field provoked me to think more carefully about the writing part of my project. I did this in the knowledge that writing not only attempts to mirror reality, but also helps to construct reality – that writing involves both *de*scription and *in*scription. I was provoked in this way because much of the writing on mobility I encountered prior to leaving for Australia did not fully prepare me for the mobilities I encountered during my empirical work. This writing included the texts by Clifford and Urry referred to at the top of this chapter – manifesto-style texts in which I read of a deterritorialised and postnational world of migrants, tourists, information and communication technologies, airports, hotels and so on. It also included books by Arjun Appadurai and Michael Peter Smith. In *Modernity at Large* (1996), Appadurai begins with two transnational processes – electronic media and mass migration – which are driven by economic opportunity, droughts, famines, leisure industries and tourist sites, and which are enabled by automobiles, aeroplanes, cameras, computers and telephones. These two transnational processes, he argues, offer new resources for the construction of imagined selves and worlds; produce diasporic public spheres, transnational or even postnational sodalities, and a sense of the global as modern and the modern as global; give rise to new patriotisms (queer nation, for example) and postnational social forms (multinational corporations, transnational philanthropic movements, international terrorist organisations etc.); and place the nation-state 'on its last legs' (p. 19). The general theory of cultural processes developed by Appadurai from this argument contains five scapes – ethnoscapes, mediascapes, technoscapes, financescapes, and ideoscapes – the relationship between which is disjunctive, since each scape is subject to its own incentives and constraints, leading to deterritorialisation.

A similar set of concerns is found in Smith's *Transnational Urbanism* (Smith, 2001). For Smith, two transnational political events – the end of the Cold War and the spread of the neoliberal variant of globalisation – displaced millions of political refugees and economic migrants during the 1990s. Meanwhile, new means of communication and travel facilitated back and forth movements of people and ideas, multi-sited projects and exchanges of material resources. As a result, he argues, we now live in transnational times, the characteristic social form of which is the migrant network. Caught up in this transnational moment, cities become places where criss-crossing transnational circuits of communication and cross-cutting transnational social practices come together in disorderly and contingent ways. Nation-states constitute and mediate flows of transnational investment, migration and cultural production across borders: sending states work to reincorporate out-migrants into their projects (as foreign investors, for example); receiving states police their borders physically through immigration legislation, and discursively through new nationalist ideologies. Smith acknowledges a constitutive and mediating role for nation-states, but, like Appadurai (and Clifford, and Urry), his agenda-setting and lightly polemical text foregrounds certain mobilities over others, so as to build his case for transnational urbanism. For example, he approaches Los Angeles through its Korean population and their transnational lives, and not through its poor, black population, most of whom are excluded from networks (Gooding-Williams, 1993; Thrift, 2002) and, like 86% of all US citizens, do not even own a passport (Hertsgaard, 2003). Similarly, Smith approaches New York through the case of Amadou Diallo, a street vendor originally from Guinea, brought to public attention when he was shot by members of the New York Police Department. The case of Diallo works in Smith's story to suggest that even African street vendors are on the move in this transnational moment. But Amadou's mother owned a gem trading company back in Guinea. She was good friends with the Guinean Foreign Minister. She sent Amadou to school in Thailand. Though not privileged in the North American context, Amadou was certainly privileged in the Guinean context.

There is no space in Smith's tale for those Guineans with no access to the internet, no satellite television, and no hope of getting an entry visa for the United States of America.

In summary, much of the writing on mobility I encountered prior to leaving for Australia encouraged me to look for, and taught me to see, migrants, tourists, electronic media, transnational projects, post-national social forms, new patriotisms etc. It should be noted that, even at the beginning of 2001, the academic literature on mobility was diverse. It ranged from the nomadism of Braidotti (1994), for whom mobility brings forth a new subject – the nomad – free from the authority of family and nation-state, to the critical mobility studies of Cresswell (2001), for whom mobilities are produced out of material conditions, and Crang (2002), for whom the non-places of super-modernity (see Augé, 1995) are characterised by hierarchy and exploitation. Beyond this literature, it should also be noted, existed a long-standing concern with the limits to mobility. Harvey (1982), for example, contrasts the hyper-mobility of money with the limited mobility of production (limited by fixed capital) and labour (limited by immigration policy, pensions policy etc.). Torpey (2000), in his biography of the passport, describes how this technology for regulating movement was invented just as many people became relatively free to move in the wake of the French Revolution and the birth of capitalism. It was used increasingly during the twentieth century, in a context of new transportation technologies, security threats, and location-specific welfare entitlements. These other texts and literatures were fascinating and highly relevant to my research problem. But they were not as widely read or acclaimed at the time as the manifesto-type writings outlined above which encouraged an exciting new focus for exciting new times. It was this excitement that took me to Australia, where a series of encounters provoked me to think more carefully about writing and mobility.

One of the first things I encountered on arriving in Australia was a political row over immigration policy. There had recently been the events now known as September 11th or 9/11. There had recently been the rape of some white women by some men of Lebanese origin in Sydney's western suburbs. Australian troops had recently stormed a Norwegian tanker to prevent the landing of 438 refugees rescued by its crew from the sea off Christmas Island. On the back of these and other events, John Howard's Liberal-National Coalition had just been returned to power in an election dominated by immigration policy, and with the words 'Australia has an absolute right as a sovereign country to decide who comes here'. During 2001–2002, immigration policy was rarely out of the news in Australia. Six detention centres had been established where around 1000 refugees were held behind razor wire for up to five years. At Woomera, the largest and most notorious of these centres, detainees sewed their lips together and went on hunger strike to protest against their treatment. On 15 May 2002, the day after Budget Day that year, Australia's *Daily Telegraph* led with the title 'Fortress Australia' above details of additional funds for border protection: AU\$77 million for Maritime Unit Surveillance; AU\$28 million for Coastwatch Air Surveillance; and AU\$13 million for Customs (a new radar system). *Yet I was there looking for deterritorialisation and postnationalism.*

Another encounter was with working holiday makers themselves. Some had felt under pressure to stay at home and progress straight from school to university, or from university to employment, or from one job to another. Some had even felt under pressure just to remain where they were because of the security offered by their current situation. For example, one interviewee who back in England had worked in local government described how "People at work just didn't get it at all. They couldn't understand why I'd leave a council job, which is supposed to be secure and quite easy, and go away and perhaps give up that security". Some working holiday makers imagined this pressure to settle into a job, and into relatively sedentary home and family life, would be there for years to come, and so they were taking the opportunity to travel now before settling down later. While some felt this expectation or norm as a suffocating pressure, others spoke longingly of a return to family and friends, and what they saw as the next life-cycle stage of settling down and building a life predominantly in one place. These working holiday makers were not convinced by the claims of new communications technologies. They found relationships difficult to maintain across distance. After receiving e-mails from family and friends, they felt like they were missing out on things back home. After receiving international telephone calls, they felt more and not less homesick. *Yet I was there looking for nomads and cosmopolitans.*

A third encounter was with residents of Sydney and their representatives. The Mayor of Sydney welcomes working holiday makers to downtown Sydney for the life they bring to the area after its office workers have returned to their suburban homes for the evening or weekend. The Mayor of Randwick, by contrast, fumes about working holiday makers in the supposedly quiet beach community of Coogee, for the late-night noise they make, the parking spaces they occupy, and the rubbish they leave behind them after moving on. The Mayor of Waverley has similar concerns about Bondi. Crucially, residents of Coogee and Bondi are not what people sometimes pejoratively describe as 'locals' – parochial people, hostile to outsiders. In the census of 1996, almost 40% of these residents reported having themselves been born overseas. What does distinguish these residents from working holiday makers, however, is their relatively sedentary lifestyles. They work regular hours to service their mortgages, take their kids to and from local schools, put the rubbish out on collection day, sleep between certain hours of the night, and find the arrivals, departures and travelling/dwelling practices of working holiday makers to be disruptive of these relatively place-based routines. *So I was there looking for mobility and finding, alongside the exemplary mobility I sought, forces fixing bodies in space through legislation and technology, desires for relatively sedentary lives or lifecycle stages, and, at the very least, pockets of shorter and slower mobility.*

Evoking mobility

These various encounters provoked a number of questions for my research project? What was this mobility I was pursuing? Was it a property of certain people and things, or was it part of and the sign for a more generalised condition? I had the impression from the literature that mobility stood for a more generalised condition. New transport and communications technologies, the end of the Cold War, and the spread of the neoliberal variant of globalisation had together generated new mobilities, which in turn were forging a new era characterised by deterritorialisation and trans- or postnationalism. If this was the case, how was mobility to be written about? How was a generalised condition to be brought to print – without reducing it to one set of practices or another, and to one set of experiences or another?

Since the research project ended, these questions have been addressed, to some extent, by a more mature mobility studies. Tim Cresswell's book-length consideration of mobility – *On the Move* (2006) – takes forward his position that mobilities are produced through forms of abstraction such as law, and surveys the highly differentiated mobility of recent and contemporary times. Two set-pieces in the book illustrate this highly differentiated mobility particularly well. The first is the chapter on Schiphol Airport. Here, the discourse of mobility rights and the micromanagement of mobility combine to produce a kinetic hierarchy of business travel, economy travel, relatively frictionless travel within Europe by European citizens (Schengen travel), illegal travel, forced travel and so on. The epilogue on Hurricane Katrina is the second set-piece that works particularly well to evoke contemporary mobility in all of its contested plurality. Cresswell recalls media images of poor black people dependent on under-funded public transport and stuck in New Orleans as the hurricane approached. He recalls alternative images of rich white people leaving the city via well-funded freeways. British tourists were rescued from deteriorating places of shelter while other categories of people were left there to ride out the storm and its aftermath. There was a debate about whether those people who did make it out should be called refugees or not. While, for some, they were clearly seeking refuge in other communities, for others, the term refugee had negative connotations inappropriate for US citizens.

A similar attempt to capture the multiple dimensions of mobility is made in the founding editorial of the journal *Mobilities* (Hannam *et al.*, 2006). The authors begin by observing how the whole world appears to be on the move, and arguing that, while mobility is not new, its speed and intensity is greater than before. But they go on to chart a course between what they term sedentarist social science (that treats the stable and proximate as normal, and, after Heidegger, focuses on dwelling) and nomadic social science (that focuses on deterritorialisation, postnationalism, and the supposed freedoms of nomadic life). From this course, mobility is viewed as dependent on the territorial concentration of resources; as organised through highly embedded and immobile material infrastructures, platforms or moorings – from airports to petrol stations. Between these nodes, routes contain fast and slow lanes, and gates that let some fluids through more than others ('the fast-tracked kinetic elite').

In this final section, I want to add something to these more nuanced framings of mobility. If what is often being described/inscribed in writings on mobility is a generalised condition or era, then we can turn to writings on other conditions and eras for examples of how best to bring such internally differentiated categories to print. Towards the end of my research project on Australia's working holiday programme, I found two such examples in the literature on modernity: Marshall Berman's *All That Is Solid Melts into Air* (Berman, 1982); and Martin Albrow's *The Global Age* (Albrow, 1996). Let us take Berman first. He is interested in the adventures, horrors, ambiguities and ironies of modern life, which he reads in texts (Goethe's Faust, The Communist Manifesto etc.), spatial and social environments (Haussmann's Parisian boulevards, Moses' New York highways etc.), and individual lives (Goethe, Marx, Baudelaire etc.). From this reading, he identifies a set of paradoxical and distinctively modern concerns – a sensibility which has cut across the boundaries of geography and society for almost 500 years: on the one hand, a will to change, transform, grow, and an attraction to knowledge and new possibilities ('the desire for development'); and, on the other hand, feelings of disorientation and disintegration, and desires for stability and coherence ('the tragedy of development'). It is by these concerns that Berman defines modernist culture. And this culture arises from the maelstrom or vortex that is fed by social world-historical processes (modernisation): discoveries in the sciences; industrialisation; demographic upheavals; urban growth; mass communication; nation-states; social movements – all born of and driven by a capitalist world market characterised by growth, waste and instability.

One need not accept that modernisation was straightforwardly born of and driven by capitalism in order to appreciate Berman's insight into representations of modernity. He notes how the first modernists – Marx, Nietzsche, Kierkegaard, Whitman, Baudelaire, Dostoevsky etc. – wrote with a particular *rhythm and range* about modernity. Their writings evoked its pace and energy, but also the broad range of practices and experiences found during the period. He then turns to the twentieth century, when, rather than engaging with the ambiguities, contradictions, ironies and tensions of modernity, writers tended increasingly towards crude, closed, flat polarities and totalisations, either embracing modernity without qualification (the Italian futurists, the Bauhaus, Le Corbusier, Marshall McLuhan, Alvin Tofler etc.) or condemning it with equal force (Max Weber, Herbert Marcuse, Michel Foucault etc.). Berman's preference is clearly for *the both/and dialectical modernism* of the nineteenth century. My encounters in the field pushed me in this direction also, towards *a both/and dialectical mobility*, that evokes both the rhythm and range of life in contemporary times.

Let us turn, finally, to Martin Albrow's exemplary treatment of the Modern Age (Albrow, 1996). This epoch, he argues, exhibited certain features (rationality, territoriality, expansion, innovation, applied science, the state, capitalism etc.), but cannot be reduced to these features. Rather, it can only be fully understood by considering all sectors, spheres, activities and events that existed and took place during the period – including religion, the economy, science, politics, disease, ideas, discoveries, revolutions, natural disasters etc. A *theoretical thrust* for the Modern Age can be identified, he argues – for him, this theoretical thrust is expanded control in space and time. But a *historical narrative* of the epoch must be a larger and more complex configuration, so that core or profiling factors are supplemented by other factors: rural populations, poor countries, traditional cultures, contingent historical events etc. Again, one need not accept this particular take on history and geography in order to appreciate Albrow's insight into the representation of generalised conditions or eras. He notes how the expansion of modernity rested on the belief that truth and error, knowledge and ignorance, rationality and irrationality take their shares from the same pool. Since the pool of human experience is expanding and infinite, however, he posits that modernity is best approached through *the rationality/irrationality tension or dichotomy*. There was no expansion of rationality at the expense of irrationality. But life was categorised using the binary code of rationality and irrationality. In my own research project, I found this insight useful. Life appears to be increasingly categorised using *the binary code of mobility and immobility*. To a large extent, a theoretical thrust for the contemporary era has been identified, along with core or profiling factors (see the literature reviewed in Sections 1 and 2 of this chapter). But *the historical narrative of mobility* is still being written. It will be complete only when it evokes the rhythm and range of provocative experiences and practices encountered by those following mobility.

9
Catching a Glimpse: The Value of Video in Evoking, Understanding and Representing the Practice of Cycling

Katrina Brown and Justin Spinney

The methodological journey to headcam video ethnography

Exploring mobile bodies *in situ*

Researchers seeking to understand how places, spaces and subjectivities are constituted in and through motion increasingly acknowledge the value of harnessing the material, social and affective context of motion itself; actively engaging movement in the creation of research knowledges. For certain highly mobile and physical practices such as cycling, attempting to research through 'being there' is crucial if we are to move beyond rationalised and decontextualised understandings of everyday mobility and explore decisions and meanings which arise in the context of movement itself. However eliciting such accounts poses particular epistemological and methodological challenges. The aim of this chapter is to share insights from recent research that has used headcam video ethnography to understand the mobile practices of others.

Drawing upon ethnographic fieldwork studying different styles of cycling in London (UK) and rural Scotland, we will both elaborate upon the particular challenges of researching highly mobile subjects, and describe the methodological journey taken to address them. In doing so we highlight the limits of established static, decontextualised and language-based methods for understanding cycling mobilities, and discuss some of the practical problems of talking and riding with cyclists. We then explain how these understandings led us to (independently) experiment with forms of mobile video ethnography where participants filmed their journeys and the subsequent video account of their cycling practices was then used as a platform for discursive exchange between researcher and participant.

By allowing riders to talk about their practices in relation to the contexts of their performance, we suggest that video can enable participants to provide more nuanced, situated and richer linguistic accounts of their embodied mobile practices than would otherwise be possible. In doing so detail, habits, sensory and emotional aspects and meanings are highlighted that might otherwise remain 'unspeakable'. Whilst not without its problems as a form of visual representation we argue here that when used in conjunction with language, the visuality of video has the potential to provide insight into aspects of mobile practices which might otherwise be neglected.

The rationalisation of practice

Cycling as an area of academic enquiry – much like the physical practice it attempts to reflect – 'enjoys' a marginal status. However, this is perhaps not so much to say that cycling has received less attention as an area of inquiry,[1] rather that particular cycling practices – and the multiplicity of meanings attached to them – have received limited attention. We argue here that leisure cycling has been marginalised in no small part because it does not fit into the category 'transport' and therefore both transport geographers and environmental activists have seen little of worth in studying such a 'diversion' from more economically and environmentally productive styles of movement. Similarly, even within styles of cycling that can be (somewhat reductively we would argue) conceived of as 'transport', a heterogeneity of meanings has been persistently passed over in favour of a narrower set of 'rational' explanatory factors such as time, distance and safety which have come to dominate the debate about why people do or do not cycle.

Certainly cycling research has been (and still is) dominated by transport geography and its associated methodological tool kit which emphasises the calculable, modellable, and predictable aspects of it as a practice (for a more detailed account of this trend see Spinney, 2009). However, there has been increasing disillusionment regarding the ability of such a narrow range of methods to explain particular aspects of the practice of cycling, and indeed other forms of movement. Whilst the orthodox instruments of transport geographers, such as stated preference surveys and traffic counts, might tell us something about the 'rational(ised)' push and pull factors of cycling, they fail to unlock the more 'unspeakable' and 'non-rational(ised)' meanings of cycling which reside in the sensory, embodied and social nature of its performance. The point that we want to make regarding much cycling research in this vein is that it is overly concerned with instrumental factors as wholly determining why and how people move around. What about the intangible and ephemeral, the sensory, emotional, kinaesthetic and symbolic aspects of cycling? Do they not play a part in defining people's experience and the meanings of cycling? As Anable and Gatersleben (2005) note, consideration of affective factors needs to be more prominent in cycling research if more realistic understandings of people's mobility and their travel choices are to be unearthed.

In contrast to previous approaches, one strand of enquiry within the mobilities turn (Cresswell, 2006; Sheller & Urry, 2006a) has begun to ask whether movement itself can be productive of meaningful associations. Certainly when we started to ask ourselves questions about other ways in which cycling becomes meaningful; about how places, meanings and identities are made and remade in and through mobile practices of cycling as forms of spatially, temporally and socially situated interaction, it became clear that the accepted tools of transport geography could offer only limited insight.

In this regard, tools such as the stated preference survey are problematic on three fronts. Firstly, such approaches assume that decisions about where, how and when to cycle are premised upon factors such as time, distance and safety with little thought to other factors which might affect the framing and weighing up of perceived costs and benefits. Secondly, when responding, such methods require participants to think through and consciously rationalise their actions and their reasons for acting in a particular way. Thirdly and related to this point, orthodox methods ask participants to articulate this rationalisation using spoken or written language. We argue that such approaches fail to take into account the fact that everyday practices are rarely rationalised in terms of a narrow range of factors, if indeed they are consciously rationalised at all. We would further argue that many of the experiences that make cycling meaningful are fleeting, ephemeral and corporeal in nature and do not lend themselves to apprehension by language alone. Indeed, if we accept that a large part of what makes people's movement meaningful to them is not rationalised at the level of language, purely linguistic accounts of movement become somewhat problematic and significantly limit any explanatory potential. Thus as Pink (2001a) asserts in relation to using video in academic research, we need to think about what kinds of knowledges we are trying to represent and how best to represent them, and employ methods accordingly (142).

In distilling life down to the page as it seems academics are destined to do, linguistic accounts all too often lose the context that makes experience meaningful, and that is the experience of being; of living experienced through the body. They also regularly place great reliance on ready-made concepts and vocabularies. Willis (2000) points out that language can be particularly poor at articulating sensuous meanings as these are generally unformed and 'in cultural solution': 'bits of meaning inhere separately in material items or liminally in elements of practice. They cannot easily be invaded or absorbed by linguistic meaning...' (27–28). However, Howes (2005) points out that due to a renewed emphasis on embodied experiences which foreground what is *done* rather than what is represented (Crang, 2003: 499; see also Ingold, 2000), the limits of language are no longer the limits of the world (2). We are not saying that such an alternative strategy avoids any rationalisation of practice; we acknowledge that all forms of representation require thought and are by their nature reductive. Rather that some strategies of eliciting data have the potential to retain or evoke some of the everyday context and detail of riding and therefore may be able to add a more nuanced and empathetic understanding to linguistic accounts. Consequently they may also help direct attention and rationalisation towards previously neglected aspects of practice that the participant, rather than the researcher, feels is important.

Vision and video

Developing the 'ride-along': exploring the limits of talking and cycling

There has in recent years been a surge of interest in techniques which emphasise 'being there' including auto-ethnography and the 'go-along' forms of mobile ethnography. In Palmer's (1996) account of French racing cyclists she draws upon observations of both her own practices and those of other participants. Palmer's main aims were to understand the embodied, social and symbolic aspects of racing and training. However, she recounts how arduous and difficult doing so was; despite being extremely fit she relates how the intensity, distance and danger of the first group research rides she went on reduced her to tears. Auto-ethnography has also been used to investigate other cycling cultures and contexts (Fincham, 2004, 2006; Jones, 2005; O'Connor & Brown, 2007; Palmer, 1996; Spinney, 2006).

Auto-ethnography does however have obvious limitations when trying to relate personal experiences to any wider group. Numerous researchers have therefore attempted to retain the 'doing' element of ethnographic research whilst seeking to understand the experiences and movements of *others*. Lee and Ingold (2006), for example, have focused on the practice of walking to demonstrate how places are created through routes, Lorimer and Lund (2003) have focused upon the technological aspects of hill-walking to position it as a technical and technological accomplishment, and Pink (2007) has used a video walk-along technique to show how people perceive the environment through the senses and constitute place through everyday mundane practice.

In these examples the 'location' of fieldwork is constantly moving yet the pace and mode of movement allows the researcher to be present. A logical extension of the walk-along therefore is to attempt to adapt the technique to cycling in the shape of the 'ride-along'. Certainly, the solution for Palmer (1996, 2001), and in Spinney's (2006) research with racing cyclists, was to ride and talk. On quiet lanes and roads with little traffic this approach is not hugely problematic as long as the researcher is fit enough to keep up; as Spinney (2006) notes, there are plenty of opportunities to talk on 4–5 hour training rides (see also O'Connor & Brown, 2007). However, the particular modalities, spatialities and pace of cycling often poses significant problems for the researcher with respect to eliciting knowledges in the context of the practice itself (Brown *et al.*, 2008; Palmer, 1996; Spinney, 2006, 2007). As Palmer (1996) has noted, the '...incessant movement between multiple sites of experience' of cyclists presents a range of problems (28).

A central issue for investigating certain mobile practices *through* those practices is how the participant can share experience or engage in communication with the researcher, whilst still allowing those practices in question to unfold. When it came to conducting research on the often solitary and traffic-laden practices of urban cycling, or the demanding rough terrain of mountain biking, our initial problem was one of how to follow people on everyday journeys, and be able to talk to them about those journeys *in the context of* those journeys. As with a number of highly mobile practices, cycling throws up a particular set of methodological challenges with regard to the speed, skill and risks associated with its practice, particularly in urban and off-road environments.

Brown's investigation of mountain bikers[2] found the ride-along to have limited utility for all but a few modes and settings of off-road cycling. On one hand, these rides enabled much of the emotional and visceral attunement and empathy that comes from such 'co-ingredience' of the researcher with informant and environment (Anderson, 2004). They also partially exploited the 'conditions, qualities and rhythms of the body in motion' in the spirit of the peripatetic tradition (Jarvis, 1997: ix), as well as the potential of the environment to foreground and situate the flow of the participant's associations and recollections of movement through particular spaces (Anderson, 2004; Kusenbach, 2003).

On the other hand, however, the nature, socialities and settings of most mountain biking practices make the simultaneous performance of the roles of mountain biker and competent researcher either imposs-ible or highly dangerous. The speed of travel and the often narrow and rugged character of the trail make proximate, sustained dialogue tricky. Moreover, the required focus of concentration often precludes the sharing of 'head-space' as well as material space. A more feasible and safe position some metres behind the participant means the researcher often misses or finds it difficult to subsequently verbalise many of the cycling 'moments' of interest, including associated but less overt emo-tions, intuitions, embodiments and expressions. The sporadic times and spaces of 're-grouping' offered opportunities for conversation between participant and researcher, but these were often dominated by wider group dynamics and the fact that we were then, strictly speak-ing, no longer 'biking'. Achieving the discursive co-presence of the method was more practically feasible for casual and family oriented riders as they tended to use flatter, wider trails, but they did not always possess the embodied skills necessary to allow safe on-the-move interviewing.

Similarly, in Spinney's fieldwork with cyclists in London[3] it was occasionally possible for more confident participants in quieter moments and places to talk about the journey being made. However in most instances this proved impossible due to the demands of riding in peak-time London traffic. In addition, there are other cycling cul-tures whose styles of riding totally preclude talking to riders about what they are doing due to the skill, intensity and solo nature of the practices they are engaged in. This is particularly true in the case of bicycle messengers (both at work and certainly whilst racing), BMX, street mountain bikers and trials riders.

Because of its inability to consistently facilitate meaningful discur-sive exchange, we found the value of the ride-along to be limited. Whilst useful in certain contexts, a ride-along is often either unsafe or risks disrupting the very practice it seeks to investigate. The more mobile, risky and skilled the practice, the more this appears to be the case. Thus, whilst there are a variety of methods being used to address some of the problems of mobile researching (Latham, 2003; Middleton, 2008; Watts & Urry, 2008), one possible way forward that we both (independently) arrived at, was to employ video as a way of retaining and evoking some of the context and detail of the practices under scrutiny whilst allowing the researcher to talk through practices 'as they happen' during playback with participants.

Video may seem like an odd choice because as a form of vision it will be positioned by many in the social sciences as suspect. Nowhere is this suspicion more evident than in geography where as Crang (2003) notes, the detached and distancing gaze has been theorised as the (masculine) antithesis of collaborative and engaged qualitative work (2003: 500, see for example Rose, 1993). Certainly Kindon (2003) suggests that, given the association of the gaze and vision in geography, it is not surprising that video has been little utilised (143). Ingold (2000), however, suggests that perhaps the association of vision and distance has more to do with the ways in which theorists have taken preconceived ideas about vision as other than active and generative into their studies (2000: 253, see also Jay, 1993, 1999; Shields, 2004). Parr (2007), for example, has used video as part of a participatory methodology to explore the worlds of those with mental health problems, and Kindon (2003) has used video as part of a participatory feminist practice of looking (see also Pink, 2006; Shrum *et al.*, 2005). Congruently there is a cautious but growing interest in video methodologies (Brown *et al.*, 2008; Downing & Tenney, 2008; Latham, 2003; Laurier, 2004; Mausner, 2008; Pink, 2001a, 2001b, 2006, 2007a; Spinney, 2007; 2009).

The value of video

Facilitating access to particular groups and practices

Perhaps the first thing that should be noted about using video is that it became a participant in the process of research; as a materiality it opened up worlds both by facilitating access and by bringing events to wider (or temporally distinct) audiences. This was particularly evident in certain cycling worlds such as trials riding, BMX and mountain biking, all of which have a strong visual culture.[4] When Spinney first met up with a group of trials riders at the Shell Centre on the South Bank he asked if he could film them. Somewhat to his surprise, rather than be seen as invasive this turned out to have been one of the best things he could have asked. By way of example, when one trials rider was asked if he could be filmed riding, he promptly got out his own camera and started talking about different techniques he used to film trials riding. So not only was filming often acceptable, in some cases it served as a tool which helped gain access to riders. It provided both insight into the particular aspects of culture that participants chose to emphasise, and a culturally congruent vehicle through which this could be done.

Similarly, Brown found the growing culture of headcam filming in mountain biking smoothed its passage into ethnographic use. Many mountain bikers were enthused to participate in the research when they learned they would get a copy of their ride on DVD, and, unlike the walkers in the same study, readily felt at ease using and wearing the equipment. Far from being alienating, the technology of the camera in these instances constitutes an object that connects people and which, as Shrum *et al.* suggest, serves to blur the boundaries between researcher and researched (2005: 17, see also Kindon, 2003; Pink, 2006; Parr, 2007). As Crang (1997) suggests, video should not be seen as simply recording the event, it can also be part of the event (366).

During the course of our research we found that many cycling practices and the feelings they engender could not be performed and talked about at the same time. Attempting a running commentary would often either disrupt the flow of a practice or leave the participant talking but no longer, strictly speaking, *in situ*. The use of headcam video in particular allows the researcher a kind of access to mobile practices which does not preclude the very happening of those practices. The technique does not require the participant to 'actively' film nor does it require the bodily and discursive presence of the researcher for filming. It therefore limits disturbance of the social, spatial and corporeal dynamic necessary to allow a particular practice to be performed and become meaningful whilst still having a presence within the research encounter. By using headcam we as researchers could 'get at', yet minimise the interruption of, important aspects of cycling practice from peak to painful experiences. The key quality of video was thus to facilitate a form of 'place-travel' or contextual borrowing or reconstruction; manipulating the time-spaces of the activity in a way that allows the participant to 're-live' it and juxtapose its audio-video representation with further layers of linguistic representation.

Defining a flexible format for incorporating video into mobile ethnography

Working video into the research process was however not a completely straightforward process.[5] In Brown's work (Brown *et al.*, 2008), the camera was mounted on the helmet of each of 24 participants on one of their routine outings, providing footage of a 'view from' perspective.

In the case of Spinney's (2008) study of urban cycling, a biographical interview was followed by filming three rides with each of 20 participants taking in different journeys/spaces in London. In both cases, various permutations of situating the researcher, participant and camera in the research encounter were tested as part of exploring the range of ways in which creating and interpreting video might become an active part of the social performances of researching and riding. For example, the headcam could be worn by either the participant or the researcher riding behind the participant.

The use of headcam video does not preclude the researcher being present either in a 'go-along' capacity or simply as an 'observant participant'. Indeed, this usually aids a deeper understanding. However, as with any method it pays to be flexible and it was not always possible or appropriate to conduct case studies in this manner. For these outings the participant was either given the video equipment beforehand to film the outing themselves, or in the case of Spinney's fieldwork a handheld video was sometimes used where the first person perspective was unworkable[6] or failed to adequately evoke the practice in question.[7] The post-outing review interviews involved both the participant and researcher taking turns to navigate through the unedited video footage, interweaving their viewing and showing with further layers of linguistic representation.

Evoking and understanding the 'unspeakable'

According to Pink (2007a) video can be used to evoke empathetic understanding in the researcher.[8] In our experience such understanding is deepened when the footage is also used by the participant to 're-live' (embodied) memories, and together researcher and participant create and negotiate new vocabularies (see for example Smith, 2002) and understandings. As a form of research data and representation, video embodies the movement which the fixity of photography and written text so often fail to evoke. Whilst not 'seeing' reality, video to an extent avoids the enemy of experience embodied in academic abstractions by relating the transient particularly well. It is therefore very useful for evoking some of the intense experiences and encounters of cycling, and getting down *into* the city or country. Video opens up movement for analysis in ways that would be impossible with a static image.

Video is unique in this regard because it continuously represents elements of space and time to produce a text, thereby opening up the possibility of analysing the rhythms of journeys; the pauses, stops, flows, weaving, waiting, rapidity and freewheeling, all of which say an awful lot about how and why people ride. In conjunction with people's accounts of what they do, such analysis allows new meanings of existing practices to be excavated. We suggest here that as we begin to illuminate how people use space and their bodies, how they interact and where and how they look, we gain a far clearer idea of how they are deriving meanings through movement and are far less likely to impose meanings upon static subjects that we are gazing at: video has the potential to bring us into the picture.

For example, one mountain biker really struggled to find words for the kinaesthetic feelings of various moments in traversing steep or rough ground but was able to use moving images to evoke and convey the emotional feelings and thoughts that accompanied those particular sequences of motions and explain how they made that (part of the) journey meaningful (see Fig. 9.1).

Video was also particularly useful in getting at feelings of flow[9] which we found to be important in both off-road and urban cycling. For example, whilst the journey to work is often conceptualised as a utilitarian act of transport, the use of video enabled riders to foreground other functions, such as riders attempting to maximise flow in the context of their journeys to and from work. One participant used the video to point to particular moments in his journey where he achieved flow, or where he failed and why. It allowed him to illustrate the micro-spatial strategies which comprised the 'ideal conditions' which Ford & Brown suggest enable flow to be achieved and experienced (159) including weaving through traffic, track-standing, riding smoothly and running red lights. Certainly the accounts of this participant had much in common with those that Ford & Brown (2006) recount in relation to surfing, communicating feelings of fluidity, gliding, being in the present, rhythmicity and unconsciousness of flow.

By using frame-by-frame video footage of participants' riding during interviews, we were thus able to elicit in-depth and 'embodied' accounts of what participants were doing and feeling. The importance of this is threefold: Firstly by getting to 'ride' with participants and asking them to talk about their movements, language is not replaced as a way of knowing, rather the video allows language to be used where it was previously difficult or impossible. The second point is that by allowing repeat and slow motion viewing, video offers a route to the other senses by stretching out the fleeting and ephemeral in order that they might be apprehended by the viewer. Buscher (2005) conceives of this functionality of video as a form of 'time-travel'. Following this line

Fig. 9.1   Mountain biker approaching and negotiating rough, steep terrain

of thinking, the first-person perspective of headcam video might also be seen as a form of micro-spatial 'travel' in that no-one (other than the participant wearing the camera) would previously have had such an impression of what it was like to 'see' from that person's body, in a certain sense 'being there' in their shoes. The final point to be made is that as a result of this process of being able to talk through bodily meanings, we begin to construct a vocabulary[10] for the unspeakable and thus language can begin to play more of a role in how we understand and represent the embodied, the momentary, the emotional and the sensual. In this sense, video becomes a bridge between embodied practice and language, enhancing the ability of language to express the ephemeral and embodied. By 'talking through' the images, expressions and language in the video footage, the meanings associated with participants' experiences were foregrounded in a co-generative way that could not have come from images or language alone.

A further strength of video for our research was the visibility it gave to the taken-for-granted, such as the mundane and ordinary rhythms, details and practices of cycling (e.g. pedalling, breathing, way-finding, drinking, re-grouping), and the sometimes-unexpected ways in which they were significant to participants. These were aspects that in go-along techniques might have been avoided as awkward 'silences', and in traditional in-depth interviews would rarely be mentioned at all. Moreover, video enabled deeper understandings of how the ordinary and extraordinary, and their affective relationships are interwoven into the 'everyday' experience of cycling. One mountain biker did not realise just how much 'unremarkable' riding was done in proportion to more 'peak' or extraordinary experiences until we talked through the footage. Using the video of his ride, he was then able to work through and verbalise how the emotional highs only existed because of the investment in – and contrast with – the emotional lows (and indeed 'moderates').

The mobility of place

The accounts of some urban riders suggested that their journeys linked together particular seemingly mundane geographical locations and memories to produce a sense of place (see Fig. 9.2). Nowhere was this more evident than with a 55 year old mother from North London. When reviewing footage of one journey on the laptop, she was able to take control of the video and forward and rewind to particular parts of the journey where she would then talk about

Fig. 9.2   The mobility of place: from the saddle

Fig. 9.2   The mobility of place: from the saddle – *continued*

particular family events and people which related to these places and made them meaningful. In doing so she linked up a number of places which collectively contributed to how she identified with London as where she belonged. Part of the rationale for her route choice beyond reasons of directness and safety was to be able to move through these places as part of a coincident process of place and identity creation. Of course these insights may just as well have been picked up if we had been able to talk as we rode. However, as this was not possible, the video facilitated a balance between being there, and being able to reflect on what had happened, ultimately leading to these insights.

Similarly, particular places of cycling are seen to be (re)made through the happenings and rememberings of pain, mishap and injury. Mountain bikers regularly explained their doings[11] in relation to particular rocks, roots, tree trunks and other such sites of past hurt, and the physical and emotional scars of which often continued to punctuate their (and other's) experiences of them. Sometimes these sites were ascribed informal names such as the 'broken ankle steps' or 'Davie's face-plant river'; serving as navigational reference points whilst, for many, colouring subsequent affective encounters.

By helping to evoke and situate past experiences, video also made visible the moments of frustration and sometimes humiliation that are part and parcel of the process of embodying skilled practice (see Fig. 9.3),[12] but which participants would often rather forget. The yelps and squeaks of excitement and fear and the grunts of bodily adjustment in relation to bike and trail that had meant little in the audio-recordings of the 'ride-alongs', now with video could be explored and 're-lived' together with the participant. Video thus created a space to explore highly embodied (and therefore taken-for-granted) competencies and tacit knowledges. Moments and sequences not seen as notable by the participant (e.g. the reflexive adjustment of balance to avoid rocks or weaving through traffic) could be opened up for exploration by the researcher by asking 'talk me through that: what did it feel like?'

Video ethnography thus helped clarify the way in which many aspects of cycling only make sense, cognitively, aesthetically and corporeally, as a mobile, processual place. For example, fundamental aspects of mountain biking only made sense to participants as motion across textured terrain – and through accompanying sensations, emotions and meanings – in the same way as Braille only makes sense by moving a finger across the raised dots on the page. For most research participants, the practice of mountain biking is focused

Fig. 9.3 Embodying skilled practice: car hanging

Fig. 9.3   Embodying skilled practice: car hanging – *continued*

Fig. 9.3 Embodying skilled practice: getting it wrong

Fig. 9.3   Embodying skilled practice: getting it wrong – *continued*

on the journey and the affective and physical relations with 'nature' enacted along the way. The destination is usually unimportant (and indeed tends to be where the journey started), and place is made and remade through grounded yet dynamic 'feelings' and seeings. Thus whilst many journeys have a destination and goal in reaching point B, they also have goals and multiple 'destinations', blurring and folding into each other, which are only found *en route* and thus place is constituted through the sensory and affective sensations of doing as well as arriving.

Conclusions

Recounting our research experience with mobile video ethnography gives a flavour of the insights that recording, viewing and talking about situated, moving visual images can contribute to the understanding of mobilities. Indeed, we argue that the situated visualities of mobile video ethnography – situated in terms of the socialities and places of movement in question and in relation to what participants have to say about their video representation – allow richer and more nuanced accounts of the mobile, embodied practices of participants than would otherwise be possible.

This holds particularly for highly mobile, physical and/or visual socio-material practices, such as cycling. Generating knowledge and understanding *of* cycling *in* and *through* the spatial and social contexts of cycling via the creation and consideration of mobile video footage provides a fresh perspective. This perspective adds crucial and sophisticated insight into the co-constituency of embodied movement and place, particularly in comparison with established techniques associated with decontextualised and 'rationalised' reflection. As we have noted, the particular transient and momentary entanglements of bodies, technologies, senses, feelings, expressions and motions with particular places that constitute the performance of cycling practices often preclude the researcher gaining 'access' to, and the participant's articulation of, these times and spaces using discursive, observational or auto-ethnographic techniques alone.

By employing mobile headcam video ethnography in particular, in a sense we are challenging what '*in situ*' means or what 'being there' (or perhaps more appropriately 'doing there') actually entails or does for the researcher and their ability to understand the experiences and meanings of others. In some cases it may be that a researcher's understanding of the mobile practices of another, based on viewing and 'talking through' a first-person dynamic visual representation of 'being there', may be deeper than if they were actually there. This is because 'talking through' video allows not only the time travel into the fleeting and ephemeral (noted by Buscher, 2005) but also 'place travel' into an intimate, dynamic, virtual 'being there'. Such virtual 'riding with' is not as a cyclist would typically ride with another, nor a simple viewing of the footage, but as an intertextual evocation of inhabiting and attaching meaning to the bodily experience of riding, with its accompanying feelings and thoughts, through the entwining of moving image and language.

By using video representations of the mobile experiences of another, we are not so much learning to see as they see (after Grasseni, 2004) but learning *through* seeing; learning what being and doing there feels like and means to participants by giving them a chance to revisit, re-live, and elaborate on these often taken-for-granted time-spaces. In short, we elicit linguistic knowledges that would have remained 'unspeakable' in a language-only research encounter. Treating vision as active and generative in this way (after Ingold, 2000) indicates the possibility of employing visualities in the creation of knowledges that are not necessarily distanciating or disempowering, thus helping to reclaim visuality as a legitimate medium through which to do social/geographical research.

Notes

1   See for example the large historical literature on cycling as a gendered, classed and racialised practice (Bailey, 1978; Baker, 1979; Burstall, 2004; Dodge, 1996; Garvey, 1995; Lewis, 1997; Mackintosh & Norcliffe, 2007; Markham, 1996; McCrone, 1991; Oddy, 2000; Oddy, 2007; Parratt, 1999, 2000; Petty, 1996, 1999, 2003; Simpson, 2003, 2007; Thompson, 2002; Vertinsky, 1991), the not inconsiderable body of largely quantitative transport geography literature on cycling (Aultman-Hall *et al.*, 1997; Axhausen *et al.*, 2000; Cervero & Radisch, 1996; Crane & Crepeau, 1998; Dill & Carr, 2003; Epperson, 1994; Forrester, 1994; Garling & Axhausen, 2003; Krizek & Johnson, 2006; Pikora *et al.*, 2003; Rietveld & Daniel, 2004; Tilahun *et al.*, 2007; Wardlaw, 2000; Wardman *et al.*, 1997), or the body of medical literature which conceives of the cycling body as a machine (Bentley *et al.*, 1998; Ericson, 1986; Kenefick *et al.*, 2002; Lucia *et al.*, 2001; Ryschon & Stray-Gundersen, 1991; Saunders *et al.*, 2005).
2   Research conducted between 2005 and 2008 in which mountain biking practices in the forests and moors of northeast Scotland were explored (in tandem with walking practices) using 'ride-along' and headcam video techniques (see Brown *et al.*, 2008; Brown, forthcoming).
3   The fieldwork was conducted between 2004 and 2006 with a range of cyclists including commuters, activists, BMX riders, messengers and trials riders.

4   Evidenced in the large range of trials riding and mountain biking footage available on TV and DVD and the growing popularity of filming by practitioners themselves.

5   Employing mobile video ethnography necessitates engagement with a range of practical, epistemological and ethical issues. For example, equipment choice, camera set up and positioning, 'framing' and situating practices, gaining access to and consent from participants, researcher fitness and skill, 'literacy' with respect to particular visual cultures, protecting the anonymity of 'incidental' participants, and breaking the law on camera, are all considerations. More detailed discussion of such issues is provided in Spinney (2008) and Brown *et al.* (2008).

6   The equipment was also prone to breakage when mounted on the bikes or person of those engaged in such 'extreme' forms of riding.

7   For example, with BMX and trials riders a handheld camera was generally used from a third person perspective because these styles are largely concerned with body/object relations.

8   It comes as little surprise then that audiencing is important to the success of such knowing and empathetic understanding (as it would be in the reception of any representation). Both Marks (2000) and Pink (2006) have emphasised a theory of audience which notes the cultural specificity of sensory experience (Marks, 2000: 195 in Pink, 2006: 53, see also Jarvinen, 2006). The question of cultural literacy in interpreting moving images is an interesting one. One could argue that video elicitation is only useful when the researcher has high literacy of a practice with a concomitant attuning of their sensibilities to likely meanings and experiences. Conversely, one could argue that video elicitation is of most use when the researcher has low cultural literacy of a practice, so that there is a platform on which they can begin to tune into the mobile worlds of others. Accordingly, a challenge faced by both authors was the potential lack of kinaesthetic empathy in relating to particular ways of doing cycling. For Spinney, his lack of experience in trials and BMX riding posed possible problems for relating to those specific styles of riding. Similarly, Brown found it harder to have immediate understanding of the embodied experiences of those riders who had skill-levels significantly above or beneath her own.

9   According to Ford & Brown (2006) flow is a form of peak experience which is characterised by a high level of confidence and control and a '...sense of rhythm and flow, with a sense of being on "auto-pilot", free and absorbed in the moment' (159).

10 Our methodological approach attempted to speak to two key audiences (at least in the first instance): participant and researcher. The former had already experienced the practice unfolding once before. The latter usually had a degree of empathetic understanding from similar (if not the same) kinds of practices. Therefore, they were able to move quite quickly to an advanced vocabulary – but may have been more likely to take the same things for granted.

11 Where doings were often punctuated by sudden immobility.

12 For an interesting and insightful account on the process of becoming and embodiment in relation to the skilled practice of Parkour see Saville, 2008.

10
Have Backpack Will Travel: Auto/biography as a Mobile Methodology

Gayle Letherby

Introduction

Influenced by the work of Charles Wright Mills, some social scientists argue for an auto/biographical approach to study. Auto/biographical work focuses on the importance of one, several or many lives and self-conscious auto/biography recognises the relationship between the self and other within the research and writing process. There is now recognition amongst social scientists that we need to consider how the researcher as author is positioned in relation to the research process: how the process affects the product in relation to the choice and design of the research fieldwork and analysis, editorship and representation. Some researchers go further and draw on their own autobiographies throughout the research and presentation process, including themselves when analysing the data and writing up. So in this type of auto/biographical writing the researcher/writer explicitly draws on their own experience as data.

In this chapter I argue that auto/biography is an illuminating methodology for studying mobility. The remainder of the chapter is divided into three sections. In 'Introducing Auto/biography' I outline the definitions and uses of auto/biography in more detail and consider its value for the study of mobility. In 'Tales from the Tracks (and the road, sea etc. etc.)' I draw on three research and writing projects to demonstrate my own use of auto/biography when studying travel and transport. In the first of these projects I, along with Gillian Reynolds (Coventry University), drew on our own experiences as well as those of 100+ respondents in our analysis of work, play and politics on the railways; the second project, also undertaken with Gillian Reynolds, involved the editing of a collection of writings (many of them auto/biographical) focusing on gender, emotion and travel and the final project with Jon Shaw (University of Plymouth) is to include auto/biographical diaries (kept by both respondents and researchers) in a study of travel, transport and respect across the generations. I conclude the chapter with some 'Final Auto/biographical Reflections'. Throughout I reflect on the value and critique of taking an auto/biographical approach.

Introducing auto/biography

Defining auto/biography

Robert Zussman (2000: 5) employs the concept of 'autobiographical occasions' to describe those moments:

> ...including job, credit and school applications, confessions both religious and criminal, reunions of various sorts, diary writing, the display of photo albums, and therapies of various sorts – at which men and women are encouraged and, at times, required to provide accounts of themselves. (Zussman, 2000: 5)

Autobiography, of course, is a first person account of a set of life experiences. Biography, on the other hand is an account of a life (or lives) written by another/a third party. Sociological autobiography involves the sociologist turning the 'sociological imagination' (Mills, 1959) towards their own lives and sociologists writing the biographies of others merge personal lives with the world of ideas (Denzin, 1989). Auto/biographical study – either focusing on one, several or many lives – highlights the need to liberate the individual from individualism; to demonstrate how individuals are social selves – which is important because a focus on the individual can contribute to the understanding of the general (Mills, 1959; Stanley, 1992; Okley, 1992; Evans, 1997). In addition auto/biographical work highlights the relationship(s) and similarities and differences between the self and other within the research and writing process. Thus, as David Morgan (1998: 655) notes:

> [auto/biography is not]...simply a shorthand representation of autobiography and/or biography but also [a] recognition of the interdependence of the two enterprises... . In writing another's life we also write or rewrite our own lives; in writing about ourselves we also construct ourselves as some body different from the person who

routinely and unproblematically inhabits and moves through social space and time.

Auto/biographical methodology

With reference to the research process it has now become common-place for the researcher to locate her/himself within the research process and produce 'first person' accounts. This clearly involves a recognition that, as researchers, we need to realise that our research activities tell us things about ourselves as well as about those we are researching (Steier, 1991; Morgan, 1998). Further, there is recognition among social scientists that we need to consider how the researcher as author is positioned in relation to the research process: how the process affects the product in relation to the choice and design of the research fieldwork and analysis, editorship and presentation (Iles, 1992; Sparkes, 1998; Letherby, 2003).

Within research, issues of auto/biography are complex. Norman Denzin and Yvonna Lincoln (1994) contend that there is frequently a false distinction made between a text and its author and they suggest that all texts bear traces of the author and are to some extent personal statements. Liz Stanley (1992: 7) observes that 'everything you produce is an auto/biography of one sort or another.' In essence every text becomes a biographical endeavour, working from the self to the other and back again, involving intersections of the public/private domains of the researcher and the researched. The use of 'I' as Stanley (1993: 49–50) notes, explicitly recognises that knowledge is contextual, situ-ational and specific, and that it will differ systematically according to the social location (as a gendered, race, classed, sexualised person) of the particular knowledge-producer. In terms of auto/biographical inclusion and connection to research some researchers go further and draw on their own autobiography throughout the research and presentation process, including them*selves* when analysing the data and writing up, which may involve inclusion of their own experience as data.

An explicitly auto/biographical approach, with its focus on the relationships between the self/other relationship also encourages reflection on power relationships within research. Zussman (2000: 6) argues one aspect of 'the narrative turn in sociology has been pre-occupied with "giving voice" to those for whom a voice has been denied'. He continues:

> ...autobiographical narratives have been taken as a way to create selves for those – most importantly women and people of color – to

whom selfhood has often been denied. Here...I am sympathetic, not just for the giving of voice itself but also for the fundamental recognition that selfhood defends on the ability to tell a story about oneself. But I am also sceptical of the assumption, implicit in much of this literature, that finding a voice is always an act of a liberating self. By calling attention to the social structures that require stories, the concept of autobiographical occasions also calls attention to the interests of those other than the autobiographer herself in the ways that stories of the self are told. (Zussman, 2000: 6)

So whilst acknowledging the political aspects of the research process within which power, emotion, involvement, detachment are all implicit it is important not to define the research process itself as political activity (Glucksmann, 1994; Letherby, 2003). In addition it is necessary to critically reflect on the political motivations of researchers and the researched and acknowledge that the imagined or potential connection between respondents and researchers is sometimes the motivation for research. Yet, it is not always possible or desirable to research issues close to us (Wilkinson & Kitzinger, 1996). Furthermore, identification should not be seen as a prerequisite to 'good' research and it is inaccurate to assume that **all** research is grounded in the autobiography of researchers. In addition, researchers do not always identify with respondents and visa versa even when they share an experience and/or identity (Letherby & Zdrodowski, 1995). Thus, researchers do not have to draw on their own life experiences to do **good** work but our life experiences/identity are present at some level in all that we do and that it is important to acknowledge this (Cotterill & Letherby, 1993; Fine, 1994; Letherby, 2003; Katz Rothman, 2007):

...we take...[respondents] autobiographies and become their biographers, while recognizing that the autobiographies we are given are influenced by the research relationship. In other words respondents have their own view of what the researcher might like to hear. Moreover, we draw on our own experiences to help us to understand those of our respondents. Thus, their lives are filtered through us and the filtered stories of our lives are present (whether we admit it or not) in our written accounts. (Cotterill & Letherby, 1993: 74)

Whether the stories we use are our own, or those of our informants, or those we cull from tables of statistically organized data, we remain story-tellers, narrators, making sense of the world as best we can... . We owe something...to our readers and to the larger community to which we offer our work. Among the many things we owe them, is an honesty about ourselves: who we are as characters in our own stories and as actors in our own research. (Katz-Rothman, 2007: unpaginated)

Given this I argue that research is always auto/biographical in that when reflecting on and writing our own autobiographies we reflect on our relationship with the biographies of others and when writing the biographies of others we inevitably refer to and reflect on our own autobiographies. Acknowledging this makes our work academically rigorous as '...self conscious auto/biographical writing acknowledges the social location of the writer thus making clear the author's role in constructing rather than discovering the story/the knowledge' (Letherby, 2000a: 90).

When I first read Mills I was impressed by his view of the social scientist's place within society: 'The social scientist is not some auto-nomous being standing outside society, the question is where he (sic) stands within it...' (Mills, 1959: 204). These strictures plus his instruc-tion to use the 'sociological imagination' as a 'tool' that enables the individual to grasp history and biography and the relations between the two seemed to me to articulate both the task and approach of sociology. I was also impressed by Mills' advice regarding the use of personal life experience in intellectual work and his view that we are personally involved in the intellectual work that we do:

...learn to use your life experience in your intellectual work: con-tinually to examine it and interpret it. In this sense craftsmanship (sic) is the centre of yourself and you are personally involved in every intellectual product which you work. (Mills, 1959: 216)

Clearly his identification of sociology/social science as auto/biographical has influenced others also, demonstrated (in the UK at least) by an extremely active Study Group focusing on Auto/Biography which although sponsored by the British Sociological Association attracts academics from a range of disciplines including history, education, occupational therapy and health.

Despite the many advantages of auto/biography and the fact that it is becoming more usual for researchers to include aspects of the self in their research and their writing, there is a tendency to keep personal details outside of the main report of a study and/or academic publication and it seems that many still feel uncomfortable with this way of writing. An understandable reason for this is the protection of oneself, one's significant others, and one's respondents (who may be more identifiable if an author writes auto/biographically) (e.g. see Scott, 1998). There is also the fear that those who write auto/biographically may be criticised for self-indulgence and weak intellectual work (Katz Rothman, 1986; Scott, 1998; Letherby, 2000). Bogusia Temple (1997: 53) suggests that 'the notion of collegial accountability to a research community is problematic'. She cites the work of Eric Mykhalovskiy (1996) whose auto/biographical writing has been described as 'self-indulgent' by an academic orthodoxy which stands by its view that there is one correct way to write about research and only one audience – a (traditional) academic audience – that know how to read 'correctly' (Mykhalovskiy, 1996; Temple, 1997, see also Sparkes, 2002 on the challenges of and to new forms of academic writing). Auto/biographical work then is a challenge to the orthodoxy and perhaps attack feels like the best form of defence.

The potential (and sometimes real) threat of professional, intellectual, and emotional attack explains why sometimes researchers and writers write about 'the personal' outside of the main report of a study. As Marion McMahon (1996: 320) states in the abstract of her article concerned with her own experience as a nonmother researching and writing about motherhood:

> This article looks at how research accounts can conceal stories about the experiences of those who do not appear to be present in the research project. Some of those who do not appear to be present may be called 'significantly absent' because their invisibility holds particular significance for the sorts of research stories researchers tell...

So, researching and writing auto/biographically is not without problems. Additionally, it is important to consider and avoid the possibility of reconstructing the past to fit the present; and the danger of privileging the voice of the researcher/writer. Thus, it is necessary to acknowledge that the power of editorship lies in the hands of the writer who must take care to respect the identity of respondents, of themselves and of their significant others (see e.g. Scott, 1998; Kirkman, 1999, 2004; Letherby, 2000). However, although the balance of power lies with the researcher/writer in terms of the production of research products/outputs, writing that is specifically auto/biographical is subject, like all research and writing, to the scrutiny of others.

As Judith Okley and Helen Callaway (1992), argue there is a fine line between 'situating oneself' and 'egotistical self-absorption". However, following Okley and Callaway (1992: 2), I would agree that a reflexive auto/biographical approach is not (as it is sometimes presented) 'navel gazing' and 'self absorption'. As Okley (1992) adds, 'self-adoration' is quite different from self-awareness and a critical scrutiny of the self. Indeed, those who protect the self from scrutiny could well be labelled self-satisfied and arrogant in presuming their presence and relations with others to be unproblematic. As Zussman (2000: 5) argues:

> Autobiographical narratives – the stories we tell about ourselves – do not simply represent the self. Neither do they simply express the self. Narratives constitute the self. This however, is on the very beginning of a serious analysis. For if autobiographical narratives constitute the self, those narratives are themselves socially constructed... . If we are to make sense of variations across time and place, then we must pay as much attention to the social structures that produce autobiographical narratives as to the narratives themselves.

Yet it is important to acknowledge that auto/biographical 'voices' within academia remain predominantly white, educated, and middle-class and Western (Bertram, 1998), so not everyone has equal access to the auto/biographical. Furthermore, auto/biography is always partial in that the writer has the power to edit the final account and that hidden selves and shadows of others are present in the writing (e.g. Iles, 1992; McMahon, 1996; Letherby, 2003).

Auto/biography and mobility

Travel is more than the concept of transport, or being transported from A to B in a box and/or on wheels. It is also more than a euphemism for tourism but less (at least in the context of this book) than an all-embracing concept of movement through spaces. A concern with travel is one aspect of the 'turn to mobilities' (Sheller & Urry, 2006; Urry, 2004, 2007) and within this: 'mobility needs to be reconsidered as a multi-layered concept, rather than the mere accumulation of miles travelled' (Fay, 2008: 65).

Laura Watts and John Urry (2008: 862) contend that the study of mobilities 'as a wide-ranging category of connection, distance and motion transforms social science and its research methods'. This could be seen to suggest the need to seek out new methods but there is also a need to highlight the implications for existing methods and methodological approaches. Auto/biography as I have described it is a methodological approach that encourages the use of particular methods and challenges traditional epistemological claims for an objective social science (e.g. Stanley, 1993; Mykhalovskiy, 1996; Letherby, 2003).

> The narrative turn moves from a singular, monolithic conception of social science toward a pluralism that promotes multiple forms of representation and research; away from facts and toward meanings; away from master narratives and toward local stories; away from idolizing categorical thought and abstracted theory and toward embracing the values of irony, emotionality, and activism; away from assuming the stance of disinterested spectator and toward assuming the posture of a feeling, embodied, and vulnerable observer; away from writing essays and toward telling stories. (Bochner, 2001, 134–135)

Tales from the tracks (and the road, sea etc. etc)

Train tracks

Gillian Reynolds and I met as Sociology undergraduates and cemented our friendship in the early 1990s as postgraduates sharing an office. Our shared interest in auto/biography is reflected in one of our earliest publications: a multi-authored book chapter focusing on experiences of the postgraduate process (Holliday *et al.*, 1993). Following the completion of our PhDs – Gillian's on disability and work and mine on the experience of 'infertility' and 'involuntary childlessness' – at Staffordshire University Gillian got a job in Birmingham and I got a post at Coventry University. We both still lived in Staffordshire and would sometimes meet on the train on the way to or from work. Our first mobilities project emerged from discussions about our current and previous love of and frustration with trains and train travel.

In our book *Train Tracks: work play and politics on the railways* (Letherby & Reynolds, 2005) we considered work, play and politics on the train and as such explored the social and cultural aspects of the train and train travel. The 'turn to mobilities' (Sheller & Urry, 2006) has led to a number of studies focusing on travel space but little had been published by social scientists on this issue when we began this project. Our aim then was to make visible an area of everyday life that we felt was largely ignored by social scientists. We drew on social and cultural theory, secondary analysis of cultural artefacts (including 'on-board' magazines, TV programmes, literature and films, and historical documents) and analysis of primary data (collected via single and focus groups interviews and email) from world-wide train travellers, workers and enthusiasts; in total over 100 respondents. Amongst other things we argued that the train has a social life in and of itself and is not simply a vehicle on wheels. We considered the personal politics of train travel and political discussion surrounding the railways, and devoted considerable space to work and leisure and the train. Thus, we aimed to demonstrate that for us and our respondents the train was a place where all social life took place. In addition we reflected on general affection for as well as frustration with trains and train travel extending Ian Marchant's (2003) analysis of the two railway systems – the railway of our imaginations and the real railway – and focused to consider both the 'nostalgic railway of our dreams' and the 'railway of consumerism and quality'. This reflected our own feelings thus:

> Our own affection and frustration with trains and train travel have intensified through data collection and writing. It has even begun to structure the gifts we buy each other: for example, a London and North Eastern Railway (LNER) mouse mat; a birthday book complete with copies of railway posters; an Australian tea towel... . (Letherby & Reynolds, 2005: 3)

Our interest in the area was clearly located in the auto/biographical as trains had been a part of each of our lives for as long as we could remember. In the introductory chapter of the book we reflected on this and the significance for our study:

> Extending the autobiographical, we are ourselves included in the respondent group, explicitly making use of the fact that all research is in some ways auto/biographical... . We have attempted a grounded analysis that, of course is influenced by our own experiences and views, both as respondents and as researchers. As such we acknowledge the intellectual and political presence of the researcher as all stages of the research process... (Letherby & Reynolds, 2005: 21)

We also 'introduced' ourselves to readers early in the book through some brief autobiographical prose, small extracts of which I reproduce here:

Gillian: Trains were an integral part of my entire childhood. The line from London to Penzance ran a couple of miles from our home village... . Thus the train became an icon of adventure and exploration for me which stays with me even as I approach my sixties... . Researching for this book was an opportunity to explore not only the train per se, but also my own love affair with it.

Gayle: As a non-car-driver, whose parents also never owned a car, the train has always been important to my life...a place where I work and play... . Sometimes I reflect on the number of days of my life I've spent on the track. But they've not been wasted, it's been productive and enjoyable... (Letherby & Reynolds, 2005: 13–14)

Throughout the book we drew on our own experience as well as those of our respondents leading to rather a mixed review of the book by Ian Carter (2007) in *The Journal of Transport History*, which demonstrates some of the complex, and arguably contradictory, critique of auto/biographical sociology:

Letherby and Reynolds bring today's conventional feminist sociological methodology to their task by foregrounding the personal, privileging qualitative methods...over quantitative, declining to swamp respondents meanings with authorial authority. This is *refreshingly different* from much previous work in railways studies... . Relentless insistence on the personal can become intrusive, of course; having watched each author expatiate in their first chapter on how she can to love Britain's railways, should we really have to *suffer* lots more long quotes from 'Gayle' and 'Gillian' (as solo arias and duets) later on? (my emphasis Carter, 2007: 131)

This critique of our auto/biographical methodology is of course not unique (see above) and auto/biographical work, like all research and writing, should be open to the scrutiny of others. Yet, arguably criticism of auto/biographical writing is harder to take as '[w]hen doing research on an issue with which one has a personal involvement and when writing in part at least about 'oneself', it is easy to feel that criticism is directed not only at your academic work but at you personally' (Letherby, 2000: 107).

Gender, travel and emotion

Gillian's and my second project in the area of mobility is an edited book (due out the week I complete the first draft of this chapter) entitled *Gendered Journeys, Mobile Emotions*. In this book the concern is with travel as it is broadly defined so although there is reference to train, bus, ship and aeroplane travel and so on there is also reference to walking, running, queuing, and to maps and campervan living. The contributors to the book include academics from various disciplines – including law, transport studies and history – who we met whilst researching and promoting 'Train Tracks' – friends (sometimes academics, sometimes not) who we knew had interesting travel tales to tell and others who we contacted via the internet (another expression of mobility Urry, 2004, 2007) following our discovery of their travel and transport related occupations and interests.

In addition to travel and transport the two concepts we asked contributors to consider in their writing were gender and emotion. Gender is still sometimes assumed to be a reflection on 'women' and 'women's issues' **but** it is equally important to consider the social and cultural expectations, behaviours and relationships of males as well as females. Thus, a thorough understanding of gender has to consider both femininities and masculinities, the range of ways in which these can be expressed and the interrelationship between gender and other signifiers of social difference (age, class, 'race', dis/ability and sexuality and so on). In addition gender is not merely something that we 'have' but rather needs to be understood more fluidly as something that is re/constructed. With specific reference to mobility and gender difference Tana Priya Uteng and Tim Cresswell (2008: 2) argue that: 'understanding [travel] mobility...means understanding observable physical movement, the meanings that such movements are encoded with, the experience of practicing these movements and the potential for undertaking these movements'.

Traditionally, mainstream sociology (and social science more generally) supported the masculinist route of 'rationality', which aimed to specifically exclude human emotion from analysis (e.g. Weber, 1968). Emotion was argued to be 'irrational', and/or the subject matter of other disciplines such as psychology. But as Simon Williams and Gillian Bendelow (1998: xvi/xvii) argue:

> ...the 'deep sociality' of emotions – offers us a way of moving beyond microanalytic, subjective, individualistic levels of analysis, towards more 'open-ended' forms of social inquiry in which embodied agency can be understood not merely as 'meaning-making', but also as 'institution making'.

Further, analysing such 'sociality' of emotions is a fundamental and necessary part of any investigation that attempts to understand, as distinct from 'rationally' categorise, the social world (Fineman, 2005). Interestingly, the word 'emotion' comes from the Latin, *emovere*, meaning 'to move, to move out' and as Sara Ahmed (2004: 11) suggests 'emotions are not only about movement, they are also about attachments or about what connects us to this or that... What moves us, what makes us feel, is also that which holds us in place, or gives us a dwelling place.' Thus:

> ...travel is an emotional experience which may promote feelings of nostalgia and affection or frustration, even anger. Trains, planes, cars and buses are spaces and places within which we may have to engage with the emotions of others... . In addition travel via motorbikes, bicycles, small single-person crafts (in the air or on water) involves engagement with other travellers in the same spaces. Travel and travelling then not only prompt emotional reactions but also at times require the management of, and work upon, one's own emotions and the emotions of others... . Emotions are also gendered, either through the differing experiences of social identities, or through cultural expectations of the 'normal' display of emotions, such as 'weeping women' or 'angry (young) men'. (Reynolds & Letherby, 2009: 22)

Gendered Journeys, Mobile Emotions contains 'traditional' academic chapters focusing on theoretical approaches, and on research concerning the experience of transport workers and travellers. It also contains 15 short autobiographical pieces. As I have done here, Gillian and I argue for the need to take auto/biographical practice within academic work seriously but for these short pieces we asked academic colleagues to 'step out' of their academic interests and disciplines and to write about issues and experiences that affected them personally, and we asked non-academic friends and acquaintances to similarly reflect on their emotional relationship to gender and travel. This meant that many authors were working outside their usual practices, outside their comfort zones, and the stories reflected the 'raw' emotions of the teller. Yet, this does not mean that there is no attention to the auto/ biographical in the rest of the book. Some contributors drew on empirical work and therefore were concerned with the biographies of others; some highlighted the significance of their identity as sociologist, historian and so on; some drew on their own autobiographical interests and experiences and some produced combinations of these. In addition, of course, this book, like most others, demanded some 'autobiographical occasions' (see Zussman, 2000 above). The 'Acknowledgements' section, gives readers some insight into our particular editorial 'debts', the introduction and final reflections chapters include some reflection on our working relationships and the 'Biographical Details' section introduces contributors always in terms of their travel and transport experiences and passions and sometimes in terms of their occupations and achievements. So auto/biographical practices of various sorts are visible throughout the book.

Travel, transport and generational respect

Job changes lead to new colleagues and new friendships. Jon Shaw and I met as colleagues at the University of Plymouth and found that we shared an interest in travel and transport. Jon, a Geographer, and I have begun to collaborate on a number of work projects including a grant proposal focusing on *Travel, Transport and Respect Across the Generations.*[1] Access to travel is uneven, often determined by car ownership/use, or the density/availability of public transport networks (Knowles *et al.*, 2008) and in turn these variables can be affected by structural and spatial factors such as income, gender, 'race' and ethnicity, dis/ability, place of residence and so on (Hine, 2008). Furthermore, perceived barriers to mobility include fear of crime and the fear/reality of anti-social activities in travel spaces such as bus and rail stations, and vehicles themselves; indeed, lack of courtesy and anti-social behaviour is recognised by transport operators and manifested in campaigns such as Transport for London's (TfL, undated) *A little thought from each of us; a big difference for everyone.* As we have argued elsewhere:

> Public transport offers different travel experiences at different times of day and in different places. The experience of travelling in the morning rush hour (full of commuters) will differ from mid-morning (shoppers), which in turn will differ from mid-afternoon (schoolchildren) and late evening (when vehicles are likely to be much emptier)... . In addition, different modes of transport are often associated with different socio-economic groups...and further complexity occurs when patterns of spatial disadvantage are factored in. Existing work has examined the travel experiences of particular age groups on particular modes...especially with regard to exclusion on account of structural or physical barriers to mobility... . Yet far less attention has been paid to issues of conflict, cohesion and associated emotional management *across* and *between* generations with

regard not only to differences of age but also to other social signifiers of difference. (Letherby & Shaw, forthcoming)

With this in mind we hope to explore existing and potential areas of cohesion, and conflict across the generations within travel spaces. This will include both an exploration of the impact of the UK government's respect agenda (Home Office, undated) and recent policy initiatives (such as those providing free and discounted travel for the over 60s and the under 18s) on changing travel habits and behaviours.

In addition to researcher involvement in 'traditional' data collection (i.e. documentary analysis of policy and newspaper reports; one-to-one interviewing and focus groups, with travellers, transport providers and policy makers, facilitated by two members of the research team), all research team members will keep their own 'travel diary' during a nine month 'travel diary' period of data collection and all will attend follow-up travel diary meetings. These diaries will be analysed along with those kept by a subgroup of our respondents. Diaries (in written, visual and audio format) have been used before by researchers of mobility. For example see *Journeys with Ada*, an audio-visual performance by Laura Watts for the UK government, Department of Transport (Watts, 2006): 'I would like to take you on a journey, a collage of the many journeys recorded as part of the research. A journey on trains, buses and foot...'.

As well as recording researchers' trajectories and corporeal experience of travel, diaries permit an active acknowledgement of the personhood of the research team members (which includes acknowledgement of physical and emotional aspects of our identity) as well as our intellectual and physical presence (Stanley & Wise, 1993; Letherby, 2003) and assist in the development of a reflexive research process. Thus, diary keeping is an auto/biographical practice:

> Diaries can provide otherwise elusive contextual details and thought processes involved in making choices and decisions and can also be used to record intentions. They provide scope for articulating issues that might not be picked up in interviews and for assessing what would otherwise remain 'hidden' or 'muted' accounts. (Cresswell & Uteng, 2008; Barbour, 2008: 18)

In preparation for our own work on travel, transport and respect across the generations we found it useful to tell each other some of our own 'Dear Diary' type 'travel stories'. Indeed, the research idea was stimulated by our early storytelling which was often about respect in transport settings in one form or other. Recently we have written a piece together (Letherby and Shaw, forthcoming) which focuses on the auto/biographical elements of our proposed research. We begin with autobiographical introductions which include:

Gayle: My relationship with higher education has always involved travel… . As my career has progressed I travel more and more **for** work, as well as **to** work. About once a month I have a meeting in London and there's usually a meeting in the midlands or the north of England twice a term. Then there are two/three sets of undergraduate or Master's external exam boards each year and two or three external PhD examinations. I probably attend on average six-eight external conferences and seminars each year, most of these in the UK but some abroad. Most years I travel to Canada, and I've been as far as Australia twice… .

Unlike most academics reflections on the travel I undertake have become part of my academic work… . My belief in the value of auto/biography and of telling stories in different sorts of ways… runs through these projects and this chapter gives me further opportunity to 'indulge' this interest.

As a non-driver public transport is also relevant to my personal as well as my academic life as I enjoy train, airplane, bus and boat travel for leisure as well as work. I admit too that although I don't drive I enjoy being driven both by friends and 'more formally': I'm the only person I know who is on first name terms with taxi drivers across the UK.

Jon: I've always been interested in travel and transport. As a kid I remember thinking that travelling by train was very civilised, not least because you could do things like read without feeling sick. As a teenager I learned to drive, and as a geography undergrad I had a yellow van in which I would cart my belongings from London to Plymouth and back… .

Back in 1995 I jumped at the chance to pursue a PhD investigating spatial aspects of British railway privatisation… . Recently I have published more and more on sustainable transport, and my studies have led me to believe that we need to reduce our demand for motorised travel, especially by car, van/lorry and aeroplane. How ironic, then, that as my career has developed I have found myself moving around ever more. …like Gayle I can use my travel to inform my professional activities. When I'm not reading or typing on the train or plane, I have taken to reflecting far more on my experiences in such a way that aids the development of projects like the one discussed in this chapter.

This project then, like the other mobility studies I have been involved in, includes auto/biographical practice at several levels.

Final auto/biographical reflections

In this chapter I have reflected on auto/biography as a useful and illuminating methodological approach within the social scientific study of mobility. Acknowledgment of this highlights, as noted above, that auto/biography is everywhere, not least in the methods we use to collect our data, in our analysis of said data and in the presentation of our findings and our theories. Adopting a position that recognises the inevitability of auto/biography encourages a critical reflexive approach that highlights the personhood of the researcher/writer, the active engagement of respondents in the process/project and the relationship between the self and other in research and academic writing:

> As reflexivity can be defined as reflecting back on something, descriptive reflexivity is clearly a description of one's reflection. Analysis means breaking something down into its constituent parts or elements and examining the relationship between them so analytical reflexivity involves comparison and evaluation. All individuals reflect on their lives and on the lives of others (Letherby, 2002: 44).

It is perhaps inevitable that I wrote some of this chapter whilst travelling on a train. This takes me back to where I began as so much of my academic work, since the beginnings of my undergraduate years, has been undertaken whilst 'on the move', not always but most often on the train. Trains (and sometimes airplanes, cars and buses) have provided me with a space to think, read and write. Although it is necessary to negotiate one's share of this public space it is certainly conducive to productive activity for me and for others; indeed I was not the only 'respondent' in the *Train Tracks* research who 'saved' certain tasks to do on the train (see Letherby & Reynolds, 2003, 2005). Furthermore, my experiences in various travel spaces have led to an area of academic work that I find both stimulating and enjoyable and which prompts excited and heated discussion with both academic and other friends.

More evidence perhaps that taking an auto/biographical approach encourages us to focus on 'real' travel stories and challenge the view that mobility is an 'elusive theoretical, social, technical and political construct' (Uteng & Cresswell, 2008: 1). With this in mind I'm off to fill my backpack in preparation for my next trip.

Note

1   We submitted this bid in collaboration with Professor Glenn Lyons, of the Centre for Transport and Society and the University of the West of England, Bristol.

Acknowledgements

My thanks to Gillian Reynolds and Jon Shaw for their friendship and for our positive working relationships. Thanks also to the editors of this volume for their helpful comments on an earlier draft.

Conclusions: Mobilising Methodologies

Mark McGuinness, Ben Fincham and Lesley Murray

Throughout the process of proposing and editing this volume we have been wary of offering any reductive or restrictive suggestions. In this spirit the following commentary is not offering any kind of final word on methodologies and mobility. Indeed this volume is intended to be a starting point in the concentrated development of distinctly mobile methodologies – tools for interrogating social relations formed through mobile practices and the sorts of contexts illustrated by the various contributors to this book.

The thinking behind *Mobile Methodologies* is rooted in a belief that the theory and practice of social research can be enhanced through methodological approaches that reflect the new mobilities agenda. In doing so it is hoped that we can establish a platform from which to encourage further experimentation and innovation in mobile research. All of our contributors have provided clear examples and suggestions for the practical application of methods to mobilities. The results of their reflection on the significance of mobility in methodological practice and research outcomes form the central thrust of the volume, namely that attending to the mobile in the method is an appropriate and often crucial exercise.

Whilst the various contributors to *Mobile Methodologies* come from a range of backgrounds and disciplines, the work included here argues that the context of movement and mobility, being *in-situ*, makes a profound difference to the sorts of things we might be able to say about the world. Each of the researchers offer insights that uphold the fundamental point that developing methodological frameworks for capitalising on an immediacy of context and the capture of *in-situ* experience, is important.

The research practices discussed throughout *Mobile Methodologies* are based on a commitment to the idea that there is a productive relationship between *research* and *context* – that research is not a passive act of collecting ready-made data waiting to be discovered and analysed by experts, but an *active* form of knowledge production.

Similarly, context is fluid, dependent on multiple, changeable variables. Attending to the productive relationship between these concepts is not without its challenges, yet each of the contributors has done so in ways that make sense in their situations and contexts. All the authors here encounter similar issues: such as how to access a subject experience that is, by definition, momentary and passing-by; issues of authenticity, of recounting corporeal, sensorial and emotional responses of participants rather than their passive observers; and issues of representation, of how to recount the mobile within the constraints of conventional academic representation and dissemination – most commonly the written word.

Each of the authors in this book are responding to an epistemological mis-fitting of social environments that are characterised by mobility but have been interrogated using research methods that have evolved out of fixed, immobile research encounters, experiences relocated or set in place to accommodate the interview, the focus group, or the questionnaire. This collection demonstrates how some researchers have attempted to resolve this problem and gives air to their reflections on research experiences.

A key motivation for compiling this collection is to highlight methodological creativity when working with mobile experience. This creative thought has been applied from various modal standpoints to demonstrate how the experience of the walker, backpacker, cyclist, car driver, passenger and so on is constituted in and through mobilities, how the corporeal and spatio-temporal contexts of mobile experiences are constitutive of personal and social meaning-making.

A consistent theme throughout the book is the importance of 'being there'. However, it is interesting to note the differing levels of commitment each of the authors has to the depth of their physical involvement in the research context. There are clear examples where authors are arguing the authenticity of their accounts of the mobile world is their research *in-situ* – Lashua and Cohen, for example, are there experiencing and witnessing alongside the research participants or involved in the research context. A very different approach is demonstrated by Laurier, who, in this volume explains his proximity to the research environment is *enabled* by using technology that facilitates his *absence*.

For him, naturalism in the research context is not impeded by the physical presence of the researcher but his ability to capture data in the moment is made possible by video technology. Despite the clear differences between these approaches the writers are attempting to get as close to meaning in data without compromising the mobile nature of its generation.

Throughout *Mobile Methodologies* authors show how they are devising methodological responses to mobile research contexts that prioritise 'being there' – however they interpret that – to understand phenomena, rather than rely on research encounters that recreate, recount or aim to recover experience that has been gained in other contexts. In this sense, each of our authors has prioritised immediacy, a first-hand social science.

As should be the case when at a starting point much of the work in *Mobile Methodologies* provokes as many questions as answers as to how best research can represent a world increasingly constituted in mobilities. For example there is the important issue of validity and trust in research. There is a question for some researchers of the extent to which a methodology based on immediacy can provide an accurate, measured account of any particular social phenomenon. [It should be noted that some might argue from a standpoint perspective that there are no such things as accurate measured accounts, just relative positions in a socio-temporal network.]

Throughout this book authors insist that social relations need to be examined in a distinct manner when considering the mobile nature of contemporary life. However, it is difficult to get a sense of which aspects of sociality are influenced by which aspects of mobility – in short it is difficult to establish the 'reach' in any given circumstance of, particularly, mundane mobilities – everyday telecommunication, everyday movement, everyday consumption, for example. It is at this point that it is worth reminding ourselves of Sheller and Urry's 'new mobilities paradigm' as offering us the opportunity to consider the discrete nature of elements of mobilities (Sheller and Urry, 2006a, 2006b).

These sorts of considerations can lead us to previously undertheorised or researched areas. Relating to the previous point an example might be the role of our impacts on others. Recognising the context of mobility as significant in the process of individuals making sense of a situation is one thing; what, however, are we to make of those whose lives are touched by the mobilities of others? The sensory world is affected by the movement of materials and people – from being stared

at by an ambling passer-by, the difficulties of crossing a heavily traf-
ficked road to go to the local shops or the resident whose cup of tea
trembles every time an intercontinental jet passes over their house.
The point is that each of these individuals has their own experiential
worlds altered and/or constrained by the mobilities of others.

The social worlds generated by the pursuit – electively or otherwise
– of unrestrained and limitless private mobility afforded through car
ownership, flight or internet connectivity have, as Laurier points out
in this collection, either been ignored or abstracted from post-facto
research out of the context of the mobile experience under investigation. Undoubtedly, we do need to understand far more about how
these mobile social worlds operate, what Urry (2007: 124) terms their
'dwellingness', to fathom who dominates, directs and decides on what
to do with these technologies of mobility.

However, what of those whose lives are touched by those moving
by without caring or noticing the impact of their mobilities? This is,
for us, a question of methodological scale. The mobile worlds of some
can (and do) undermine the quality of the worlds inhabited by others.
Those who live next to ever-busier roads, motorways, airports and
railway stations – or people that are prey to the 'victimless' crime per-
petuated through the virtual realm see their quality of life deteriorate
each time a distant other passes by, over, through or virtually enters
and leaves their social world. These are not new concerns. Doreen Massey
in her classic essay on a Global Sense of Place (Massey, 1994) acknow-
ledged the impact one person's seemingly independent decision to travel
by a particular route and transport mode might have on less powerful
others:

> This concerns not merely the issue of who moves and who doesn't,
> although that is an important element of it; it is also about power in
> relation to the flows and the movement. Different social groups
> have distinct relationships to this anyway-differentiated mobility;
> some are more in charge of it than others; some initiate flows
> and movement, others don't; some are more on the receiving end
> of it than others; some are effectively imprisoned by it. (Massey,
> 1994: 149)

This volume is intended to provoke a response from the research
community. It is very much a point of departure from which we hope
others will draw ideas and inspiration to develop new and exciting
ways of interrogating the social world. As the mobile fix becomes more

sophisticated so must our sensitivity to the influence of mobilities on our social selves – this level of sensitivity can only be achieved through appropriately sensitive research tools. In heralding in the 'new mobilities paradigm' Sheller and Urry have, by implication set the challenge for the development of a new methodological paradigm. This book demonstrates the work of researchers attempting to meet this challenge.

Bibliography

Aagaard Nielsen, K. & Svensson, L. (eds) (2006) *Action Research and Interactive Research* (Maastricht: Shaker Publishing).

Adams, J. (1999) *The Social Implications of Hypermobility* (Paris: OECD).

Adey, P. (2006) 'If mobility is everything then it is nothing: towards a relational politics of (im)mobilities', *Mobilities*, 1/1, 75–94.

Adey, P. (2007) 'May I have your attention: airport geographies of spectatorship, position and (im)mobility', *Environment and Planning D: Society and Space*, 25/3, 515–536.

Ahmed, S. (2004) *The Cultural Politics of Emotion* (Edinburgh: University of Edinburgh Press).

Albrow, M. (1996) *The Global Age: State and Society Beyond Modernity* (Cambridge: Polity).

Allen, J. (2003) *Lost Geographies of Power* (Oxford: Blackwell).

Almeida, B. (1986) *Capoeira: A Brazilian Art Form* (Berkeley: North Atlantic Books).

Alvesson, M. & Sköldberg, K. (2000) *Reflexive Methodology* (London: Sage Publications).

Amin, A. (2004) 'Regions unbound: towards a new politics of place', *Geografiska Annaler B*, 86/1, 33–44.

Anable, J. & Gatersleben, B. (2005) 'All work and no play? The work of instrumental and affective factors in work and leisure journeys by different travel modes', *Transportation Research Part A*, 39/2–3, 163–181.

Anderson, B. (2004) 'Time-stilled space-slowed: how boredom matters', *Geoforum*, 35/6, 739–754.

Anderson, J. (2004) 'Talking whilst walking: a geographical archaeology of knowledge', *Area*, 36/3, 254–261.

Appadurai, A. (1996) *Modernity at Large: Cultural Dimensions of Globalisation* (Minneapolis: University of Minnesota Press).

Assuncao, M.R. (2005) *Capoeira: The History of an Afro-Brazilian Martial Art* (London: Routledge).

Augé, M. (1995) *Non-Places: Introduction to an Anthropology of Supermodernity* (London: Verso).

Augé, M. (2002) *In the Metro* (Minneapolis: University of Minnesota Press).

Aultman-Hall, L., Hall, F.L. & Baetz, B. (1997) 'Analysis of bicycle commuter routes using geographic information systems: implications for bicycle planning', *Transportation Research Record*, 1578, 102–110.

Axhausen, K., Zimmerman, A., Schonfelder, S., Rindsfuser, G. & Haupt, T. (2000) 'Observing the rhythms of daily life: a six-week travel diary', *Transportation*, 29, 95–124.

Bærenholdt, J., Haldrup, M., Larsen, J. & Urry, J. (2004) *Performing Tourist Places* (Aldershot, Hants: Ashgate).

Bailey, P. (1978) *Leisure and Class in Victorian England: Rational Recreation and the Contest for Control, 1830–1885* (London: Routledge & Kegan Paul).

Baker, W. (1979) 'The leisure revolution in Victorian England: a review of recent literature', *Journal of Sport History*, 6/3, 76–85.

Banks, M. (2001) *Visual Methods in Social Research* (London: Sage).

Barbour, R. (2008) *Introducing Qualitative Research: A Student Guide to the Craft of Doing Qualitative Research* (London: Sage).

Basford, L., Reid, S., Lester, T., Thomson, J. & Tolmie, A. (2002) 'Drivers' perceptions of cyclists'. *Transport Research Laboratory Report*, TRL549.

Bauman, Z. (2000) *Liquid Modernity* (Cambridge: Polity Press).

Bauman, Z. (2001) *Community: Seeking Safety in an Insecure World* (Cambridge: Polity).

Bayart, J-F. (2008) *Global Subjects: A Political Critique of Globalization* (Cambridge: Polity).

Beck, U. (1992) *Risk Society: Towards a New Modernity* (London: Sage).

Becker, H. (2001) 'Georges Perec's experiments in social description', *Ethnography*, 2/1, 63–76.

Bennett, A. (2002) 'Researching youth culture and popular music: a methodological critique', *British Journal of Sociology*, 53/3, 451–466.

Bentley, D., Wilson, G., Davie, A. & Zhou, S. (1998) 'Correlations between peak power output, muscular strength and cycle time trial performance in triathletes', *Journal of Sports Medicine and Physical Fitness*, 38/3, 201–207.

Berman, M. (1982) *All That Is Solid Melts into Air: The Experience of Modernity* (New York: Simon and Schuster).

Bertram, V. (1998) 'Theorising the personal: using autobiography in academic writing', in Jackson, S. & Jones, G. (eds) *Contemporary Feminist Theories* (Edinburgh: Edinburgh University).

Bissell, D. (2007) 'Animating suspension: waiting for mobilities', *Mobilities*, 2/2, 277–298.

Bissell, D. (2010) 'Vulnerable quiescence: mobile timespaces of sleep', *Cultural Geographies*, 17/1, forthcoming.

Bissell, D. & Fuller, G. (2009) 'The revenge of still', *M/C Journal*, 12(1), http://journal.media-culture.org.au/index.php/mcjournal/ article/viewArticle/ 136.

Blaikie, N. (1993) *Approaches to Social Enquiry* (Cambridge: Polity Press).

Bochner, A. (2001) 'Narrative's Virtues', *Qualitative Inquiry*, 7/2, 131–157.

Bolland, T. (2006) *Plug Inn (The Forgotten Years)* (Liverpool: Self-published).

Bournes, D.A. & Mitchell, G.J. (2002) 'Waiting: the experiences of persons in a critical care waiting room', *Research in Nursing & Health*, 25/1, 58–67.

Braidotti, R. (1994) *Nomadic Subjects: Embodiment and Sexual Difference in Contemporary Feminist Theory* (New York: Columbia University Press).

Brekke, J-P. (2004) 'While we are waiting: uncertainty and empowerment among asylum-seekers in Sweden', *Report for Institute for Social Research*, Oslo.

Brown, K.M., Dilley, R. & Marshall, K. (2008) 'Using a head-mounted video camera to understand social worlds and experiences', *Sociological Research Online*, 13/6, http://www.socresonline.org.uk/13/6/1.html.

Browning, B. (1995) *Samba* (Bloomington, IND: Indiana University Press).

Browning, B. (1998) *Infectious Rhythm* (New York: Routledge).

Brunt, L. (1999) 'Thinking about ethnography', *Journal of Contemporary Ethnography*, 28/5, 500–509.

Bryman, A. (2004) *Social Research Methods* (Oxford: Oxford University Press).

Buetow, S. (2004) 'Patient experience of time duration: strategies for "slowing time"' and "accelerating time" in general practices', *Journal of Evaluation in Clinical Practice*, 10/1, 21–25.

Burawoy, M. (1998) 'The extended case method', *Sociological Theory*, 16(1), 4–33.

Burstall, P. (2004) *The Golden Age of the Bicycle: The Worldwide Story of Cycling in the 1890s* (Marlow on Thames: Little Croft Press).

Buscher, M. (2005) 'Social Life Under the Microscope?' *Sociological Research Online*, 10/1, <http://www.socresonline.org.uk/10/1/buscher.html>.

Buscher, M. & Urry, J. (2009) 'Mobile Methods and the Empirical', *European Journal of Social Theory*, 12/1, 99–116.

Caird, J.K., Willness, C.R., Steel, P. & Scialfa, C. (2008) 'A meta-analysis of the effects of cell phones on driver performance', *Accident Analysis and Prevention*, 40, 1282–1293.

Campos de Rosario, C., Stephens, N. & Delamont, S. (2009) 'I'm your teacher! I'm Brazilian!', *Sport, Education and Society* (forthcoming).

Canzler, W., Kaufmann, V. & Kesselring, S. (eds) (2008) *Tracing Mobilities: Towards a Cosmopolitan Perspective* (Aldershot: Ashgate).

Capoeira, N. (1995) *The Little Book of Capoeira* (Berkeley: North Atlantic Books).

Capoeira, N. (2002) *Capoeira: Roots of the Dance – Fight – Game* (Berkeley: North Atlantic Press).

Capoeira, N. (2006) *A Street-Smart Song* (Berkeley, CA: Blue Snake Books).

Carter, I. (2007) 'Review of train tracks: work, play and politics on the railways', *Journal of Transport History*, 28/1, 131–133.

Castells, M. (1996) *Information Age: Economy, Society and Culture: Rise of the Network Society, Vol 1* (Oxford: Blackwell).

Cavell, S. (1990) *Conditions Handsome and Unhandsome: The Constitution of Emersonian Perfectionism* (Chicago: University of Chicago Press).

Cavell, S. (1994) *A Pitch of Philosophy, Autobiographical Exercises* (London: Harvard University Press).

Cavell, S. (2002) *Must We Mean What We Say* (Updated Edition) (Cambridge: Cambridge University Press).

CeMoRe (2006) 'Air-time spaces' workshop at Centre for Mobilities Research, Lancaster University, 29–30 September, 2006.

Cervero, R. & Radisch, C. (1996) 'Travel choices in pedestrian versus automobile oriented neighbourhoods', *Transport Policy*, 3/3, 127–141.

Chambers, I. (1985) *Urban Rhythms: Pop Music and Popular Culture* (London: Macmillan).

Church, A., Frost, M. & Sullivan, K. (2000) 'Transport and social exclusion in London', *Transport Policy*, 7, 195–205.

Clarke, N. (2004a) 'Mobility, fixity, agency: Australia's Working Holiday Programme', *Population, Space and Place*, 10(5), 411–420.

Clarke, N. (2004b) 'Free independent travellers? British working holiday makers in Australia', *Transactions of the Institute of British Geographers*, 29/4, 499–509.

Clarke, N. (2005) 'Detailing transnational lives of the middle: British working holiday makers in Australia', *Journal of Ethnic and Migration Studies*, 31/2, 307–322.

Clifford, J. (1992) 'Travelling cultures', in Grossburg, L. (ed.) *Cultural Studies*, 96–116.

Clifford, J. (1997) *Routes: Travel and Translation in the Late Twentieth Century* (London: Harvard University Press).

Cohen, S. (1993) 'Ethnography and popular music studies', *Popular Music*, 12/2, 123–138.

Cohen, S. (2007) *Decline, Renewal and the City in Popular Music Culture: Beyond the Beatles* (Aldershot: Ashgate).

Conradson, D. (2005) 'Freedom, space and perspective: moving encounters with other ecologies', in Davidson, J., Bondi, L. & Smith, M. (eds) *Emotional Geographies* (Aldershot: Ashgate).

Cook, I. & Crang, M. (1995) *Doing Ethnographies* (London: IBG).

Corbridge, S. (1989) 'Marxism, post-Marxism, and the geography of development', in Peet, R. & Thrift, N. (eds) *New Models in Geography*, 1, 224–254.

Cotterill, P. & Letherby, G. (1993) 'Weaving stories: personal auto/biographies in feminist research', *Sociology*, 27/1, 67–79.

Crane, R. & Crepeau, R. (1998) 'Does neighbourhood design influence travel?: A behavioural analysis of travel diary and GIS data', *Transport Research*, 3/4, 225–238.

Crang, M. (2005) 'Qualitative methods: there is nothing outside the text?', *Progress in Human Geography*, 29(2), 225–233.

Crang, M. (2003) 'Qualitative methods: touchy, feely, look-see?', *Progress in Human Geography*, 27/4, 494–504.

Crang, M. (2002) 'Between places: producing hubs, flows and networks', *Environment and Planning A*, 34, 569–574.

Crang, M. (1997) 'Picturing practices: research through the tourist gaze', *Progress in Human Geography*, 21/3, 359–373.

Crary, J. (1999) *Suspensions of Perception: Attention, Spectacle, and Modern Culture* (Cambridge, MA: MIT Press).

Cresswell, T. (2001) 'The production of mobilities', *New Formations*, 43, 11–25.

Cresswell, T. (2003) *Mobilizing Place, Placing Mobility: The Politics of Representation in a Globalized World* (Amsterdam: Rodopi).

Cresswell, T. (2006) *On the Move: Mobility in the Modern Western World* (London: Routledge).

Cresswell, T. & Uteng, T.P. (2008) 'Gendered mobilities towards an holistic understanding', in Uteng, T.P. & Cresswell, T. (eds) *Gendered Mobilities* (Aldershot: Ashgate).

Daniels, S. (2006) 'Suburban pastoral: Strawberry Fields forever and sixties memory', *Cultural Geographies*, 13, 28–54.

Dant, T. (2004) 'Recording the "habitus"', in C. Pole (ed.) *Seeing is Believing? Approaches to Visual Research* (Oxford: Elsevier).

de Certeau, M. (1984) *The Practice of Everyday Life* (Berkeley and Los Angeles: University of California Press).

Delanty, G. (1997) *Social Science: Beyond Constructivism and Realism* (Buckingham: Open University Press).

Deleuze, G. & Guattari, F. (1987) *A Thousand Plateaus: Capitalism and Schizophrenia* (Minneapolis: University of Minnesota Press).

Denzin, N.K. (1989) *Interpretive Biography, Qualitative Research Methods Series* (London, New Delhi: Sage).

Denzin, N.K. & Lincoln, Y.S. (1994) *Handbook of Qualitative Methods* (London: Sage).

Denzin, N.K. & Lincoln, Y.S. (2000) *Handbook of Qualitative Research* (California: Sage Publications, Inc.).

Denzin, N.K. & Lincoln, Y.S. (2003) *Collecting and Interpreting Qualitative Materials* (Thousand Oaks: Sage).

Dill, J. & Carr, T. (2003) 'Bicycle commuting and facilities in major US cities: if you build them, commuters will use them', *Transportation Research Record*, 1828, 116–123.

Dingus, T.A., Klauer, S.G., Neale, V.L., Petersen, A., Lee, S.E., Sudweeks, J., Perez, M.A., Hankey, J., Ramsey, D., Gupta, S., Bucher, C., Doerzaph, Z.R. & Jermeland, J. (2004) *The 100-car Naturalistic Driving Study; Phase II – Results of the 100-car Field Experiment*. Contract No. DTNH22-00-C-07007 (Task Order No. 06) (Blacksburg, VA: Virginia Tech Transportation Institute).

Diski, J. (2006) *On Trying to Keep Still* (London: Little, Brown).

Dodge P. (1996) *The Bicycle* (Paris-New York: Flammarion).

Domosh, M. & Morin, K. (2003) 'Travels with feminist historical geography', *Gender, Place and Culture*, 10/3, 257–264.

Downey, G. (2005) *Learning Capoeira* (New York: Oxford University Press).

Downing, M. & Tenney, L. (eds) (2008) *Video Visions: Changing the Culture of Social Science Research* (Cambridge Scholars: Newcastle Upon Tyne).

Drewes Nielsen, L. & Aagaard Nielsen, K. (2006) Er fleksibiliteten bæredygtig? Tidsskrift for arbejdsliv, 2, 24.

Drewes Nielsen, L. (2001) 'Virksomhedsanalysen og den refleksive metode – en kritisk analyse af fleksibilitet, magt og tidspres', in Pedersen, B. & Drewes Nielsen, L. (eds) Kualitative metoder – fra metateori til markarbejde (Roskilde: Roskilde Universitetsforlag).

Drewes Nielsen, L. (2005) 'Reflexive mobility – A critical and action oriented perspective on transport research', in Thomsen, T.U., L. Drewes Nielsen & H. Gudmundsson (eds) *Social Perspectives on Mobility* (Aldershot: Ashgate).

Drewes Nielsen, L. (2006) 'The methods and implications of action research', in Aagaard, N. & Svensson, L. (eds) *Action Research and Interactive Research* (Maastricht: Shaker Publishing).

Duffy, M. (2000) 'Lines of drift: festival participation and performing a sense of place', *Popular Music*, 19(1), 51–64.

Edensor, T. (2003) 'M6: Junction 19-16: defamiliarising the mundane roadscape', *Space and Culture*, 6/2, 151–168.

Edensor, T. & Holloway, J. (2008) 'Rhythmanalysing the coach tour: the Ring of Kerry, Ireland', *Transactions of the Institute of British Geographers*, 33/4,483–501.

Emmel, N. & Clark, A. (2007) 'Learning to use visual methodologies in our research: a dialogue between two researchers', ESRC Research Development Initiative: Building capacity in visual methods, Introduction to visual methods workshop', 22–23 January 2007, University of Leeds, Leeds.

Epperson, B. (1994) 'Evaluating suitability of roadways for bicycle use: toward a cycling level-of-service standard: bicycles and bicycle facilities', *Transportation Research Record*, 1438, 9–16.

Ericson, M. (1986) 'On the biomechanics of cycling: a study of joint and muscle load during exercise on the bicycle ergometer', *Scandinavian Journal of Rehabilitation Medicine*, 16, 1–43.

Evans, M. (1997) *Introducing Contemporary Feminist Thought* (Cambridge: Polity Press).

Fay, M. (2008) '"Mobile Belonging": exploring transnational feminist theory and online connectivity', in Uteng, T.P. & Cresswell, T. (eds) *Gendered Mobilities* (Aldershot: Ashgate).

Featherstone, M., Thrift, N. & Urry, J. (eds) (2004) *Automobilities* (London: Sage).

Fielding, N. (1994) 'Ethnography', in Gilbert, M. (ed.) *Researching Social Life* (London: Sage).

Fincham, B. (2007) 'Generally speaking people are in it for the cycling and the beer: bicycle couriers, subculture and enjoyment, *Sociological Review*, 55/2, 189–202.

Fincham, B. (2007) 'Bicycle messengers: image, identity and community', in Horton, D., Rosen, P. & Cox, P. (2007) *Cycling and Society* (Aldershot: Ashgate).

Fincham, B. (2006) 'Back to the "old school": bicycle messengers, employment and ethnography', *Qualitative Research*, 6/2, 187–205.

Fincham, B. (2004) *Bicycle Couriers: Identity, Risk & Work* (Unpublished PhD Thesis, University of Cardiff).

Fine, M. (1994) 'Dis-tance and other stances: negotiations of power inside feminist research', in Gitlin, A. (ed.) *Power and Method: Political Activism and Educational Research* (London: Routledge).

Fineman, S. (2005) 'Appreciating emotion at work: paradigm tensions', *International Journal of Work, Organisation and Emotion*, 1/1, 4–19.

Finnegan, R. (1989) *The Hidden Musicians: Music-making in an English Town* (Cambridge: Cambridge University Press).

Flick, U. (2006) *An Introduction to Qualitative Research* (London: Sage).

Ford, N. & Brown, D. (2006) *Surfing and Social Theory: Experience, Embodiment and Narrative of the Dream Glide* (London: Routledge).

Forrester, J. (1994) *Bicycle Transportation: A Handbook for Cycling Transportation Engineers* (Cambridge Massachusetts: MIT Press).

Freudendal-Pedersen, M. (2009) *Mobility in Daily Life – Between Freedom and Unfreedom* (Aldershot: Ashgate).

Freudendal-Pedersen, M. (2007) 'Mobility, motility and freedom: the structural story as analytical tool for understanding the interconnection', *Swiss Journal of Sociology*, Vol. 33, No. 1, pp. 27–43.

Freudendal-Pedersen, M. & Hartmann-Petersen, K. (2006) 'Fællesskaber som udgangspunkt? Refleksiv mobilitet og human security i mobilitetsforskningen', *Nordisk Samhällsgeografisk tidsskrift*. Nr. 41/42, 175–196. Roskilde.

Freudendal-Pedersen, M., Hartmann-Petersen, K. & Roslind, K. (2002) Strukturelle fortællinger om mobilitet. Final Thesis, Roskilde University.

Furedi, F. (2008) *Paranoid Parenting* (London: Continuum Press).

Gane, N. (2006) 'Speed up or slow down? Social theory in the information age', *Information, Communication & Society*, 9/1, 20–38.

Gans, H.J. (1999) 'Participant observation in the Era of "ethnography"', *Journal of Contemporary Ethnography*, 28, 540–548.

Garfinkel, H. (1967) *Studies in Ethnomethodology* (Englewood Cliffs, NJ: Prentice-Hall).

Garling, T. & Axhausen, K. (2003) 'Introduction: habitual travel choice', *Transportation*, 30, 1–11.

Garvey, E.G. (1995) 'Reframing the bicycle: advertising-Supported Magazines and scorching women', *American Quarterly*, 47, 1, 66–101.

Geertz, C. (1973) *The Interpretation of Cultures* (New York: Basic Books).

Geertz, C. (1988) *Works and Lives – The Anthropologist as Author* (Stanford CA: Stanford University Press).

Giddens, A. (1984) *The Constitution of Society: Outline of a Theory of Structuration* (Cambridge: Polity).

Giddens, A. (1991) *Modernity and Self-Identity: Self and Society in the Late Modern Age* (Cambridge: Polity).

Gillett, C. (1983) *The Sound of the City*, 2nd Edition (London: Souvenir).

Gilmore, A. (2004) 'Popular music, urban regeneration and cultural quarters: the case of the Rope Walks, Liverpool', in D. Bell & M. Jayne (eds) *City of Quarters: Urban Villages in the Contemporary City* (Aldershot: Ashgate).

Glucksmann, M. (1994) 'The work of knowledge and the knowledge of women's work', in M. Maynard & J. Purvis (eds) *Researching Women's Lives from a Feminist Perspective* (London: Taylor and Francis).

Gooding-Williams, R. (1993) *Reading Rodney King, Reading Urban Uprising* (New York: Routledge).

Gottdiener, M. (2000) *Life in the Air: Surviving the New Culture of Air Travel* (Boulder: Rowman & Littlefield Publishers).

Graham, S. & Marvin, S. (2001) *Splintering Urbanism: Networked Infrastructure, Technological Mobilities and the Urban Condition* (London: Routledge).

Grasseni, C. (2004) 'Video and ethnographic knowledge: skilled vision and the practice of breeding', in Pink, S., Kurti, L. & Afonso, A.I. (eds) *Working Images* (London: Routledge).

Halkier, B. (2008) *Fokusgrupper* (Frederiksberg: Samfundslitteratur).

Hall, T., Lashua, B. & Coffey, A. (2006) 'Stories as sorties', *Qualitative Researcher*, 3, 2–4.

Hannam, K., Sheller, M. & Urry, J. (2006) 'Editorial: mobilities, immobilities and moorings', *Mobilities*, 1/1, 1–22.

Harley, R. (2009) 'Light-air-portals: visual notes on differential mobility', *M/C Journal*, 12(1), http://journal.media-culture.org.au/index.php/mcjournal/article/viewArticle/132.

Harrison, P. (2000) 'Making sense: embodiment and the sensibilities of the everyday', *Environment and Planning D: Society and Space*, 18/4, 497–517.

Harrison, P. (2008) 'Corporeal remains. Vulnerability, proximity, and living-on after the end of the world', *Environment and Planning A*, 42/2, 423–445.

Harrison, P. (2009) 'Remaining still', *M/C Journal*, 12(1), http://journal.media-culture.org.au/index.php/mcjournal/article/viewArticle/135.

Hartmann-Petersen, K., Freudendal-Pedersen, M. & Drewes Nielsen, L. (2007) 'Mobilitetens optik på det moderne liv', in Jensen & Andersen & Hansen & Aagaard Nielsen (eds) *Planlægning i teori og praksis – et tværfagligt perspektiv* (Roskilde: Roskilde Universitetsforlag).

Harvey, D. (1982) *The Limits to Capital* (Oxford: Blackwell).

Hertsgaard, M. (2003) *The Eagle's Shadow: Why America Fascinates and Infuriates the World* (London: Bloomsbury).

Hinchliffe, S. & Whatmore, S. (2006) 'Living cities: towards a politics of conviviality', *Science as Culture*, 15/2, 123–138.

Hine, J. (2008) 'Transport and social justice', in R. Knowles, J. Shaw & I. Docherty (eds) *Transport Geographies: Mobilities, Flows and Spaces* (Oxford: Blackwell).

Hine, J. & Mitchell, F. (2003) *Transport Disadvantage and Social Exclusion: Exclusion Mechanisms in Transport in Urban Scotland* (Aldershot: Ashgate).

Holliday, R. (2000) 'We've been framed: visualising methodology', *Sociological Review*, 48/4, 503–520.

Holliday, R., Letherby, G., Mann, L., Ramsay, K. & Reynolds, G. (1993) 'Room of Our Own: an alternative to academic isolation', in Kennedy, M., Lubelska, C. & Walsh, V. (eds) *Making Connections: Women's Studies, Women's Movements, Women's Lives* (London: Falmer Press, Taylor and Francis).

Holloway, S. & Valentine, G. (2000) *Children's Geographies: Playing, Living, Learning* (Abingdon, Oxon: Routledge).

Homan, S. (2003) *The Mayor's a Square: Live Music and Law and Order in Sydney* (Sydney: Local Consumption Publications).

Home Office (undated) Tackling anti-social behaviour and its causes. http://www.respect.gov.uk/ Accessed 29 July 2008.

Howes, D. (ed.) (2005) *Empire of the Senses: The Sensual Culture Reader* (Oxford: Berg).

Hutchinson, S. (2000) 'Waiting for the bus', *Social Text*, 18/2, 107–120.

Iles, T. (ed.) (1992) *All Sides of the Subject: Women and Biography* (New York: Teacher's College).

Ingold, T. (2000) *The Perception of the Environment: Essays in Livelihood, Dwelling and Enskillment* (London: Routledge).

Jacobsen-Hardy, M. (2002) 'Behind the razor wire: a photographic essay', *Ethnography*, 3/4, 398–415.

Jarvinen, H. (2006) 'Kinaesthesia, synaesthesia and le Sacre du Printemps: responses to dance modernism', *Senses & Society*, 1/1, 71–91.

Jarvis, R. (1997) *Romantic Writing and Pedestrian Travel* (Basingstoke: Macmillan).

Jay, M. (1993) *Downcast Eyes: The Denigration of Vision in Twentieth Century French Thought* (Berkeley: University of California Press).

Jay, M. (1999) 'Returning the gaze: the American response to the French critique of occularcentrism', in Weiss, G. & Haber, H. (eds) *Perspectives on Embodiment* (London: Routledge).

Jensen, A., Andersen, J., Hansen, O.E. & Aagaard Nielsen, K. (eds) (2007) Planlægning i teori og praksis – et tværfagligt perspektiv (Frederiksberg: Roskilde Universitetsforlag).

Jones, P. (2005) 'Performing the city: a body and a bicycle take on Birmingham, UK', *Social and Cultural Geography*, 6/6, 813–818.

Jungk, R. & Müllert, N.R. (1984) Håndbog i Fremtidsværksteder. København: Politisk Revy (Translated from: Zukunftswerkstatten, Wege zur Wiederbelebung der Demokratie (1981)).

Kaplan, C. (1996) *Questions of Travel* (Durham, NC: Duke University Press).

Katz Rothman, B. (1986) *The Tentative Pregnancy: Prenatal Diagnosis and the Future of Motherhood* (New York: Viking).

Katz Rothman, B. (2007) 'Writing ourselves in sociology', *Methodological Innovations Online*, 2(1), unpaginated.

Kaufmann, V. (2002) *Re-thinking Mobility and Contemporary Sociology* (Aldershot: Ashgate).

Kenefick, R., Mattern, C., Mahood, N. & Quinn, T. (2002) 'Physiological variables at lactate threshold under-represent cycling time-trial intensity', *Journal of Sports Medicine and Physical Fitness*, Vol. 42, No. 4, 396–402.

Kenyon, S. (2006) 'Reshaping patterns of mobility and exclusion? The impact of virtual mobility upon accessibility, mobility and social exclusion', in Sheller, M. & Urry, J. (eds) *Mobile Technologies of the City* (London: Routledge).

Kindon, S. (2003) 'Participatory video in geographic research: a feminist practice of looking?', *Area*, 35/2, 142–153.

Kirkman, M. (1999) 'I didn't interview myself: the researcher as participant in narrative research', *Annual Review of Health Social Sciences*, 9/1, 32–41.

Kirkman, M. (2004) 'Genetic connection and relationships in narratives of donor-assisted conception', *Australian Journal of Emerging Technologies and Society*, 2/1, 1–20.

Knowles, R., Shaw, J. & Docherty, I. (eds) (2008) *Transport Geographies: Mobilities, Flows and Spaces* (Oxford: Blackwell).

Koslowsky, M., Kluger, A. & Reich, M. (1995) *Commuting Stress: Causes, Effects, and Methods of Coping* (London: Plenum Press).

Kraftl, P. & Horton, J. (2008) 'Spaces of every-night life: for geographies of sleep, sleeping and sleepiness', *Progress in Human Geography*, 32/4, 509–524.

Krims, A.P. (2007) *Music and Urban Geography* (New York: Routledge).

Krizek, J. & Johnson, P. (2006) 'Proximity to trails and retail: effects on urban cycling and walking', *Journal of the American Planning Association*, Vol. 72, No. 1, 33–42.

Kuhn, A. (ed.) (1990) *Alien Zone* (London: Verso).

Kuppers, P. (2000) 'Toward the unknown body: stillness, silence, and space in mental health settings', *Theatre Topics*, 10(2), 129–143.

Kusenbach, M. (2003) 'Street phenomenology: the go-along as ethnographic research tool', *Ethnography*, 4/3, 455–485.

Kvale, S. (1996) *Interviews: An Introduction to Qualitative Research Interviewing* (Thousand Oaks: Sage).

Laing, D. (2008) 'Gigographies'. International Association for the Study of Popular Music (IASPM) United Kingdom and Ireland branch – Conference Proceedings. 13–15 September, University of Glasgow, Glasgow.

Lambert, D. (2005) 'Producing/contesting whiteness: rebellion, anti-slavery and enslavement in Barbados, 1816', *Geoforum*, 36, 29–43.

Lashua, B., Hall, T. & Coffey, A. (2006) 'Soundwalking as research method', Royal Geographical Society with the Institute of British Geographers Annual International Conference. London.

Latham, A. (2003) 'Research, performance, and doing human geography: some reflections on the diary-photography, diary-interview method', *Environment and Planning A*, 35/11, 1993–2017.

Latour, B. (1992) 'Where are the missing masses? the sociology of a few mundane artefacts', in Bijker, W.E. & Law, J. (ed.) *Shaping Technology/Building Society* (London: MIT Press).

Laurier, E. (forthcoming) 'Cognition and driving', in Merriman, P. & Cresswell, T. (eds) *Mobilities: Practices, Spaces, Subjects* (London: Ashgate).

Laurier, E. (2004) 'Doing office work on the motorway', *Theory, Culture & Society*, 21/4–5, 261–277.

Laurier, E. (2005) 'Searching for a parking space', *Intellectica*, 2–3(41/42), 101–116.

Laurier, E. & Philo, C. (2006) 'Cold shoulders and napkins handed: gestures of responsibility', *Transactions of the Institute of British Geographers*, 31/2, 193–207.

Laurier, E., Lorimer, H., Brown, B., Jones, O., Juhlin, O., Noble, A., Perry, M., Pica, D., Sormani, P., Strebel, I., Swan, L., Taylor, A.S., Watts, L. & Weilenmann, A. (2008) 'Driving and passengering: notes on the ordinary organisation of car travel', *Mobilities*, 3/1, 1–23.

Law, J. & Hassard, J. (1999) *Actor Network Theory and After* (Oxford: Blackwell).

Law, J. & Mol, A. (2001) 'Situating technoscience: an inquiry into spatialities', *Environment and Planning D: Society and Space*, 19/5, 609–621.

Lee, J. & Ingold, T. (2006) 'Fieldwork on foot: perceiving, routing, socialising', in S. Coleman & P. Collins (eds) *Locating the Field: Space, Place and Context in Anthropology*. ASA Monograph (Oxford: Berg).

Letherby, G. (2000a) 'Working and Wishing: an auto/biography of managing home and work', *Auto/Biography VIII*, 1 & 2, 89–98.

Letherby, G. (2000b) 'Dangerous Liaisons: auto/biography in research and research writing', in G. Lee-Treweek & S. Linkogle (eds) *Danger in the Field: Risk and Ethics in Social Research* (London: Routledge).

Letherby, G. (2002) 'Claims and disclaimers: knowledge, reflexivity and representation in feminist research', *Sociological Research Online*, 6:4, www.socresonline.org.uk/8/4/

Letherby, G. (2003) *Feminist Research in Theory and Practice* (Buckingham: Open University).

Letherby, G. & Reynolds, G. (2003) 'Making connections: the relationship between train travel and the process of work and leisure', *Sociological Research Online*, 8:3, http://www.socresonline.org.uk/8/3/letherby.html

Letherby, G. & Reynolds, G. (2005) *Train Tracks: Work, Play and Politics on the Railways* (London: Berg).

Letherby, G. & Reynolds, G. (eds) (2009) *Gendered Journeys, Mobile Emotions* (Aldershot: Ashgate).

Letherby, G. & Shaw, J. (forthcoming) 'Dear Diary: auto/biography, respect and mobility', in Vanninni, P. (ed.) *Routes Less Travelled: The Cultures of Alternative Mobilities* (Aldershot: Ashgate).

Letherby, G. & Zdrodowksi, D. (1995) '"Dear Researcher": the use of correspondence as a method with feminist qualitative research', *Gender and Society*, 9/5, 576–593.

Lewis, G. (1997) 'Sport, youth culture and conventionality 1920–1970', *Journal of Sport History*, 24, 1, 129–150.

Lewis, J.L. (1992) *Ring of Liberation* (Chicago: The University of Chicago Press).

Leyshon, A., Matless, D. & Revill, G. (1998) *The Place of Music* (Guilford: Routledge).

Lisle, D. (2009) 'The "Potential Mobilities" of Photography', *M/C Journal*, 12(1), http://journal.media-culture.org.au/index.php/mcjournal/article/viewArticle/125.

Livingston, E. (1995) *An Anthropology of Reading* (Bloomington and Indianapolis: Indiana University Press).

Lomax, H. & Casey, N. (1998) 'Recording social life: reflexivity and video methodology', *Sociological Research Online*, 3, U3–U32.

Lorimer, H. (2002) 'Pedestrian geographies: walking, knowing and placing Scotland's mountains', Abderdeen & Swindon, Departments of Geography and Anthropology, University of Aberdeen, ESRC, Swindon.

Lorimer, H. (2005) 'Cultural geography: the busyness of being more-than-representational', *Progress in Human Geography*, 29/1, 83–94.

Lorimer, H. (2007) 'Cultural geography: worldly shapes, differently arranged', *Progress in Human Geography*, 31, 1, 89–100.

Lorimer, H. & Lund, K. (2003) 'Performing facts: finding a way over Scotland's mountains', *Sociological Review*, 51/2, 130–144.

Lucas, K. (2004) *Running on Empty* (Bristol: The Policy Press).

Lucia, A. Hovos, J. & Chicarro, J. (2001) 'Physiology of professional road racing', *Sports Medicine*, 31/5, 325–337.

Lynch, M. (1993) *Scientific Practice and Ordinary Action: Ethnomethodology and Social Studies of Science* (Cambridge: Cambridge University Press).

Mackintosh, B. & Norcliffe, G. (2007) 'Men, women and the bicycle: gender and social geography of cycling in the late nineteenth century', in Horton, D., Rosen, P. & Cox, P. (eds) *Cycling & Society* (Aldershot UK: Ashgate).

Mallot, J. (2006) 'Body politics and the body politic', Interventions: *International Journal of Postcolonial Studies*, 8/2, 165–177.

Marchant, I. (2003) *Parallel Lines: Or, Journeys on the Railway of Dreams* (London: Bloomsbury).

Marcus, G.E. (1995) 'Ethnography in/of the world system: the emergence of multi-sited ethnography', *Annual Review of Anthropology*, 24, 95–105.

Marcus, G.E. (1998) *Ethnography Through Thick and Thin* (Princeton, NJ: Princeton University Press).

Markham, S. (1996) 'Nick Kaufmann: on a wheel against time', in Van der Plas (ed.) *Proceedings, 7th International Cycle History Conference* (Anaheim CA: KNI).

Marks, L. (2000) *The Skin of the Film* (Durham: Duke University Press).

Massey, D. (2006) *Is The World Really Shrinking?* A Festival of Ideas for the Future – Open University Radio Lecture, Thursday 9 November 2006, BBC Radio 3.

Massey, D. (2000) 'Travelling thoughts', in Gilroy, P., Grossberg, L. & McRobbie, A. (eds) *Without Guarantees: In Honour of Stuart Hall*, pp. 225–232 (London: Verso).

Massey, D. (1994) 'A global sense of place', in D. Massey (ed.) *Space, Place and Gender* (Oxford: Blackwell).

Massey, D. (1993) 'Power-geometry and a progressive sense of place', in Bird, J., Curtis, B., Putnam, T., Robertson, G. & Tickner, L. (eds) *Mapping the Futures: Local Cultures, Global Change* (London: Routledge).

Maus, M. (2006) *Techniques, Technology and Civilization* (Oxford: Berghahn).

Mausner, C. (2008) 'Capturing the hike experience on video: an alternative framework for studying human response to nature', in Downing, M. & Tenney, L. (eds) *Video Visions: Changing the Culture of Social Science Research* (Cambridge Scholars: Newcastle Upon Tyne).

May, J. & Thrift, N. (eds) (2001) *Timespace: Geographies of Temporality* (London: Routledge).

McCormack, D. (2002) 'A paper with an interest in rhythm', *Geoforum*, 33/4, 469–485.

McCormack, D. (2003) 'An event of geographical ethics in spaces of affect', *Transactions of the Institute of British Geographers*, 28/4, 488–507.

McCrone, K. (1991) 'Class, gender, and English women's sport, c. 1890–1914', *Journal of Sport History*, 18/1, 159–182.

McMahon, M. (1996) 'Significant absences', *Qualitative Inquiry*, 2/3, 320–336.

Middleton, J. (2008) 'London: the walkable city', in Imrie, R., Lees, L. & Raco, M. (eds) *Regenerating London: Governance, Sustainability and Community in a Global City* (London: Routledge).

Mills, C.W. (1959) *The Sociological Imagination* (London: Penguin).

Morgan, D. (1998) 'Sociological imaginations and imagining sociologies: bodies, auto/biographies and other mysteries', *Sociology*, 32/4, 647–663.

Mulhall, S. (2002) *On Film* (London: Routledge).

Murphie, A. (2009) 'Be still, be good, be cool: the ambivalent powers of stillness in an overactive world', *M/C Journal*, 12(1), http://journal.mediaculture.org.au/index.php/mcjournal/article/viewArticle/133.

Murray, L. (2009a) 'Making the journey to school: the gendered and generational aspects of risk in constructing everyday mobility', *Health, Risk & Society*, 11(5), 471–486.

Murray, L. (2009b) 'Looking at and looking back: visualization in mobile research', *Qualitative Research*, 9(4), 469–488.

Mykhalovskiy, E. (1996) 'Reconsidering table talk: critical thoughts on the relationship between sociology, autobiography and self-indulgence', *Qualitative Sociology*, 19(1), 131–151.

Nash, C. (2000) 'Performativity in practice: some recent work in cultural geography', *Progress in Human Geography*, 24/4, 653–664.

Nielsen, L.D. & Pedersen, K.B. (eds) (2001) Kvalitative metoder – fra metateori til markarbejde (Frederiksberg: Roskilde Universitetsforlag).

Nisbit, M. (1992) *Atget's Seven Albums* (New Haven: Yale University Press).

Normark, D. (2006) Enacting mobility: studies into the nature of road-related social interaction', Goteborg, Science and Technology Studies, University of Gothenburg.

O'Connor, J. & Brown, T. (2007) 'Real cyclists don't race: informal affiliations of the weekend warrior', *International Review for the Sociology of Sport*, 42/1, 83–97.

Oddy, N. (2000) 'Cycling in the drawing room', in Ritchie, A. & van der Plas, R. (eds) *Proceedings, 11th International Cycle History Conference* (San Francisco: Van der Plas).

Oddy, N. (2007) 'The flaneur on wheels?', in Horton, D., Rosen, P. & Cox, P. (eds) *Cycling & Society* (Aldershot UK: Ashgate).

Okley, J. (1992) 'Anthropology and autobiography: participatory experience and embodied knowledge', in Okley, J. & Callaway, H. (eds) *Anthropology and Autobiography* (London: Routledge).

Okley, J. & Callaway, H. (eds) (1992) *Anthropology and Autobiography* (London: Routledge).

Ong, A. (1999) *Flexible Citizenship: The Cultural Logics of Transnationality* (London: Duke University Press).

Palmer, C. (1996) *A Life of It's Own: The Social Construction of the Tour de France* (University of Adelaide, unpublished PhD Thesis).

Palmer, C. (2001) 'Shit happens: the selling of risk in extreme sport', in *The Australian Journal of Anthropology*, 13/3, 323–336.

Parr, H. (2007) 'Collaborative film-making as process, method and text in mental health research', *Cultural Geographies*, 14, 114–138.

Parratt, C. (1999) 'Making leisure work: women's rational recreation in late Victorian and Edwardian England', *Journal of Sport History*, 26/3, 471–487.

Parratt, C. (2000) 'Of place and men and women: gender and topophilia in the "Haxey Hood"', *Journal of Sport History*, 27/2, 229–245.

Petty, R. (1996) 'Women and the wheel: the bicycle's impact on women', in Van der Plas (ed.) *Proceedings, 7th International Cycle History Conference* (Anaheim CA: KNI).

Picart, C. (1997) 'Metaphysics in Gaston Bachelard's "reverie"', *Human Studies*, 20, 59–73.

Pikora, T., Giles-Corti, B., Bull, F., Jamrozik, K. & Donovan, R. (2003) 'Developing a framework for assessment of the environmental determinants of walking and cycling', *Social Science & Medicine*, 56, 1693–1703.

Pile, S. & Thrift, N. (eds) (2000) *Citz A-Z* (London: Routledge).

Pink, S. (2001a) *Doing Visual Ethnography* (London: Sage Publications).

Pink, S. (2001b) 'More visualising, more methodologies: on video, reflexivity and qualitative research', *Sociological Review*, 49/4, 586–599.

Pink, S. (2006) *The Future of Visual Anthropology: Engaging the Senses* (Oxon: Routledge).

Pink, S. (2007a) *Doing Visual Ethnography: Images, Media and Representation in Research* (2nd edition) (London: Sage).

Pink, S. (2007b) 'Walking with Video', *Visual Studies*, 22/3, 240–252.

Raffel, S. (1979) *Matters of Fact* (London: Routledge & Kegan Paul).

Raje, F. (2004) *Transport Demand Management and Social Inclusion: The Need for Ethnic Perspectives* (Aldershot, Hants: Ashgate).

Reason, P. & Bradbury, H. (eds) (2008) *The Sage Handbook of Action Research: Participative Inquiry and Practice* (London: Sage).

Reis, A.L.T. (2003) *Subjective Well-being of Capoeira as Physical Activity*. Unpublished PhD Thesis, University of Bristol, UK.

Reis, A.L.T. (2005) *Capoeira: Health and Social Well-being* (Brasilia: Thesaurus Editor de Brasilia Ltd.).

Reynolds, S. (2005) *Rip It Up and Start Again: Post-punk 1978–1984* (London: Faber and Faber).

Reynolds, G. & Letherby, G. (2009) 'Emotion, gender and travel: moving on', in Letherby, G. & Reynolds, G. (eds) *Gendered Journeys, Mobile Emotions* (London: Ashgate).

RGS-IBG (2006) 'Mobile methodologies', conference session at RGS-IBG Annual Conference, 30 August–1 September 2006, London.

RGS-IBG (2007) 'Animating geographies', conference session at RGS-IBG Annual Conference, 29–31 August 2007, London.

Rietveld, P. & Daniel, V. (2004) 'Determinants of bicycle use: do municipal policies matter?', *Transportation Research Part A*, 38, 531–550.

Ritchie, A. (1999) 'The origins of bicycle racing in england: technology, entertainment, sponsorship and advertising in the early history of the sport', *Journal of Sport History*, 26/3, 489–520.

Ritchie, A. (2003) 'The league of American Wheelmen, Major Taylor and the "Colour Question" in the United States in the 1890s', *Sport in Society*, 6/2–3, 13–43.

Rose, G. (1993) *Feminism and Geography: The Limits of Geographical Knowledge* (Cambridge: Polity Press).

Rose, G. (2001) *Visual Methodologies* (London: Sage).

Rose, G. (2007) *Visual Methodologies: An Introduction to the Interpretation of Visual Materials* (second edition) (London: Sage).

Rouncefield, M. (2002) '"Business as usual": an ethnography of everyday (bank) work', *Computing* (Lancaster: University of Lancaster).

Ryschon, T. & Stray-Gundersen, J. (1991) 'The effect of body position on the energy cost of cycling', *Medicine & Science in Sports & Exercise*, 23/8, 949–953.

Sacks, H. (1978) 'Some technical considerations of a dirty joke', in Schenkein, J. (ed.) *Studies in the Organization of Conversational Interaction* (New York: Academic Press).

Sacks, H. (1984) 'On doing being ordinary', in Atkinson, J.M. & Heritage, J.C. (eds) *Structures of Social Action* (Cambridge: Cambridge University Press).

Sacks, H. (1992) *Lectures on Conversation, Vol. 1* (Oxford: Blackwell).

Sacks, H., Schegloff, E.A. & Jefferson, G. (1974) 'A simplest systematics for the organization of turn-taking for conversation', *Language*, 50, 696–735.

Sanders, C.R. (1999) 'Prospects for a post-modern ethnography', *Journal of Contemporary Ethnography*, 28, 669–675.

Saunders, A., Dugas, J., Tucker, R., Lambert, M. & Noakes, T. (2005) 'The effects of different air velocities on heat storage and body temperature in humans cycling in a hot, humid environment', *Acta Physiologica Scandinavica*, 183/3, 241–255.

Saville, S. (2008) 'Playing with fear: parkour and the mobility of emotion', *Social and Cultural Geography*, 9/8, 891–914.

Schegloff, E.A. (2007) 'A tutorial on membership categorization', *Journal of Pragmatics, 39*, 462–482.

Scott, S. (1998) 'Here be dragons', *Sociological Research Online*, 3(3), www.socres-online.org.uk/socresonline/3/3/1html.

Sennett, R. (1998) *The Corrosion of Character – The Personal Consequences of Work in the New Capitalism* (New York: Norton).

Sheller, M. & Urry, J. (eds) (2006a) 'Materialities and mobilities', Special issue of *Environment and Planning A*, 38/2.

Sheller, M. & Urry, J. (eds) (2006) *Mobile Technologies of the City* (London: Routledge).

Sheller, M. & Urry, J. (2006) 'The new mobilities paradigm', *Environment and Planning*, 38/2, 207–226.

Shields, R. (2004) 'Visualicity', *Visual Cultures in Britain*, 5/1, 23–36.

Shrum, W., Duque, R. & Brown, T. (2005) 'Digital video as research practice: methodology for the millennium', *Journal of Research Practice*, 1/1, <http://www.icaap.org>

Silverman, D. (2005) *Doing Qualitative Research: A Practical Handbook* (London: Sage).

Simpson, C. (2007) 'Capitalising on curiosity: women's professional bicycle racing in the late nineteenth century', in Horton, D., Rosen, P. & Cox, P. (eds) *Cycling & Society* (Aldershot: Ashgate).

Simpson, C. (2003) 'Managing public impressions: strategies of nineteenth century female cyclists', *Sporting Traditions*, Vol. 19, No. 2, pp. 1–15.

Smith, M. (2001) *Transnational Urbanism: Locating Globalisation* (Oxford: Blackwell).

Smith, M.L. (2002) 'Moving self: the thread that bridges dance and theatre', *Research in Dance Education*, 3/2, 123–141.

Smith, T. (1995) 'Soundings: travel and tiredness', *British Medical Journal*, 311, 1441.

Spark, M. (2006) 'A neoliberal nexus: economy, security and the biopolitics of citizenship on the border', *Political Geography*, 25, 151–180.

Sparkes, A. (1998) 'Reciprocity in critical research? Some unsettling thoughts', in Shaklock, G. & Smyth, J. (eds) *Being Reflexive in Critical Educational and Social Research* (London: Falmer).

Sparkes, A. (2002) *Telling Tales in Sport and Physical Activity: A Qualitative Journey* (Leeds: Human Kinetics).

Spinney, J. (2009) 'Cycling the city: movement, meaning and method', *Geography Compass* (Blackwell online publication).

Spinney, J. (2006) 'A place of sense: a kinaesthetic ethnography of cyclists on Mont Ventoux', *Environment and Planning D: Society and Space*, 24/5, 709–732.

Spinney, J. (2007) 'Cycling the city: non-place and the sensory construction of meaning in a mobile practice', in D. Horton, P. Rosen & P. Cox (eds) *Cycling and Society* (Aldershot: Ashgate).

Spinney, J. (2008) Cycling the city: movement, meaning and practice (unpublished PhD Thesis, Royal Holloway University of London).

Spurr, D. (1993) *The Rhetoric of Empire* (Durham, NC: Duke University Press).

Stanley, L. (1993) 'On Auto/biography in sociology', *Sociology*, 27, 1, 41–52.

Stanley, L. (1992) The auto/biographical i: the theory and practice of feminist auto/biography (Manchester: Manchester University Press).

Stanley, L. & Wise, S. (1993) *Breaking Out Again: Feminist Ontology and Epistemology* (London: Routledge).

Steier, F. (1991) *Research and Reflexivity* (London: Sage).

Steiger, R. (2000) 'Enroute: an interpretation through images', *Visual Sociology*, 15, CD-ROM.

Stephens, N. & Delamont, S. (2006a) 'Balancing the berimbau', *Qualitative Inquiry*, 12/2, 316–339.

Stephens, N. & Delamont, S. (2006b) 'Samba no mar', in D. Waskul & P. Vaninni (eds) *Body/Embodiment* (Aldershot: Ashgate).

Stephens, N. & Delamont, S. (2008) 'Up on the roof', *Cultural Sociology*, 2/1, 57–74.

Stephens, N. & Delamont, S. (2010a) 'They start to get malicia', *British Journal of Sociology of Education* (forthcoming).

Stephens, N. & Delamont, S. (2010b) 'Roda Boa, Roda Boa', *Teaching and Teacher Education*, 29 (forthcoming).

Stephens, N. & Delamont, S. (2010c) *Will it Hurt?* Unpublished paper.

Temple, B. (1997) 'Collegiate accountability' and bias: the solution to the problem?', *Sociological Research Online*, 2/4, www.socresonline.org.uk/socresonline/2/4/8.

Thompson, C. (2002) 'Bicycling, class, and the politics of leisure in Belle Epoque France', in Koshar, R. (ed.) *Histories of Leisure* (Oxford: Berg).

Thomsen, T.U., Drewes Nielsen, L. & Gudmundsson, H. (eds) (2005) *Social Perspectives on Mobility* (Aldershot: Ashgate).

Thornton, S. (1995) *Club Cultures: Music, Media and Subcultural Capital* (Cambridge: Polity Press).

Thrift, N. (1996) *Spatial Formations* (London: Sage).

Thrift, N. (2002) 'A hyperactive world?', in Johnson, R., Taylor, P. & Watts, M. (eds) *Geographies of Global Change*, 2nd Edition (Oxford: Blackwell).

Thrift, N. (2004) 'Intensities of feeling: towards a spatial politics of affect', *Geografiska Annaler B*, 86/1, 57–58.

Thrift, N. (2006) 'Affecting Relations', paper presented at Association of American Geographers Annual Conference, March 2006, Chicago.

Thrift, N. (2008) *Non-representational Theory: Space, Politics, Affect* (London: Routledge).

Tilahun, N., Levinson, D. & Krizek, K. (2007) 'Trails, lanes, or traffic: valuing bicycle facilities with an adapted stated preference survey', *Transportation Research Part A*, 41, 287–301.

Torpey, J. (2000) *The Invention of the Passport: Surveillance, Citizenship and the State* (Cambridge: Cambridge University Press).

Transport for London (undated) A *Little Thought from Each of Us; a Big Difference for Everyone* (London: TfL). http://www.tfl.gov.uk/corporate/media/newscentre/ 7337.aspx Accessed 26 May 2008.

Trezise, R. (2007) *Dial M for Merthyr* (Aberteifi: Parthian Books).

Turchi, P. (2004) *Maps of the Imagination: The Writer as Cartographer* (San Antonio: Trinity University Press).

Urry, J. (2000) *Sociology Beyond Societies: Mobilities for the Twenty-First Century* (London: Routledge).

Urry, J. (2003) *Global Complexity* (Cambridge: Polity Press).

Urry, J. (2004) 'Connections', *Environment and Planning D: Society and Space*, 22, 27–37.

Urry, J. (2007) *Mobilities* (Cambridge: Polity).

Uteng, T. & Cresswell, T. (2008) *Gendered Mobilities* (Aldershot: Ashgate).

Valentine, G. (2004) *Public Space and the Culture of Childhood* (Aldershot: Ashgate).

Vertinsky, P. (1991) 'Old age, gender and physical activity: the biomedicalization of aging', *Journal of Sport History*, 8/1, 64–80.

Virilio, P. (1977) *Speed and Politics: An Essay on Dromology* (New York: Semiotext(e)).

Walker, I. (2007) 'Drivers overtaking bicyclists: objective data on the effects of riding position, helmet use, vehicle type and apparent gender', *Accident Analysis and Prevention*, 39, 417–425.

Wardlaw, M. (2000) 'Three lessons for a better cycling future', *British Medical Journal*, 321/23–30, 1582–1585.

Wardman, M., Hatfield, R. & Page, M. (1997) 'The UK national cycling strategy: can improved facilities meet the targets?', *Transport Policy*, 4/2, 123–133.

Watts, L. (2006) 'Travel times (or journeys with Ada)', Department of Sociology, Lancaster University htpp://eprints.lancs/ac.uk/4348/

Watts, L. (2008) 'The art and craft of train travel', *Social and Cultural Geography*, 9/6, 711–726.

Watts, L. (forthcoming) 'The art and craft of train travel', *Journal of Social and Cultural Geography*, Department of Sociology, Lancaster.

Watts, L. & Urry, J. (2008) 'Moving methods, travelling times', *Environment and Planning: Society and Space*, 26(5), 860–874.

Weber, M. (1988) *Economy and Society* (Edited by Guenther Roth and Claus Wittich) (New York: Bedminister Press).

Weilenmann, A. (2003) 'Doing mobility', Department of Informatics (Göteborg: Göteborg University).

Wilkinson, C. & Kitzinger, S. (eds) (1996) *Representing the Other: A Feminism and Psychology Reader* (London: Sage).

Williams, S. & Bendelow, G. (1998) 'Introduction: emotions in social life. Mapping the terrain', in G. Bendelow & S. Williams (eds) *Emotions in Social Life: Critical Themes and Contemporary Issues* (London: Routledge).

Williams, S.J. (2002) 'Sleep and health: sociological reflections on the dormant society', *Health*, 6(2), 173–200.

Williams, S.J. (2005) *Sleep and Society: Sociological Explorations and Agendas* (London: Routledge).

Williams, S.J. & Boden, S. (2004) 'Consumed with sleep? Dormant bodies in consumer culture', *Sociological Research Online*, 9, http://www.socresonline. org.uk/9/2/williams.html.

Willis, P. (2000) *The Ethnographic Imagination* (Cambridge: Polity Press).

Wylie, J. (2002) 'An essay on ascending Glastonbury Tor', *Geoforum*, 33/4, 441–454.

Wylie, J. (2005) 'A single day's walking: narrating self and landscape on the South West Coast Path', *Transactions of the Institute of British Geographers*, 30/2, 234–247.

Wylie, J. (2006) 'Smoothlands: fragments/landscapes/fragments', *Cultural Geographies*, 13/3, 458–465.

Zussman, R. (2000) Autobiographical occasions: introduction to the special i ssue qualitative sociology – Special Issue: Autobiographical Occasions, 23/1, 5–8.

Index